DATE DUE

Unsolved Mysteries
of Science

Unsolved Mysteries
of Science

A Mind-Expanding Journey
through a Universe of
Big Bangs, Particle Waves,
and Other Perplexing Concepts

JOHN MALONE

John Wiley & Sons, Inc.
New York • Chichester • Weinheim • Brisbane • Toronto • Singapore

ISBN 0-471-38441-0

Printed in the United States of America

10 9 8 7 6 5 4 3 2 1

Contents

Acknowledgments

I would like to thank my editor, Jeff Golick, for his encouragement, helpfulness, and patience throughout the writing of this book, and my agent, Bert Holtje, for bringing us together in the first place. My thanks to Paul Baldwin, Rob Brock, Dan Tepper, and Carole Monferdini for their willingness to listen to me talk endlessly about quarks, dinosaurs, and multiple universes. I am indebted to Tom Tirado for his expertise in many matters, from computers to Mayan civilization. Finally, I would like to pay tribute to John W. Campbell, whose magazine *Astounding Science Fiction* opened my eyes to whole new universes when I was a teenager in the 1950s and sparked an interest in science, which ultimately led to this book.

Introduction

Scientists, from Aristotle onward, have always seen it as their job to explain the world, to unravel its mysteries. It often seems that for every mystery solved, however, a new one is created. What's more, even the greatest scientists are capable of dealing only with some aspects of any given puzzle, and the solutions they arrive at quite often eventually turn out to have been wrong, for that very reason. Aristotle virtually created the scientific method in Western thought, but his concept of the heavens, with its crystal spheres revolving around the Earth, was about as wrong as it is possible to be. Sir Isaac Newton, who was the first to explain gravity and light in ways that truly worked in terms of the observable world, had his apple cart driven into a temporary ditch when Albert Einstein's Relativity Express roared by at the beginning of the twentieth century. Newton has had a kind of revenge, however—his demonstrable gravitational effects have resisted all attempts to integrate them into quantum physics.

Throughout most of the history of science, there had been a tendency to regard the latest theoretical triumph or technical breakthrough as the last word on the subject. At the end of the nineteenth century, there was a widespread feeling, even among scientists, that just about everything that could be discovered and explained had been addressed. Then, in the first five years of the twentieth century, human beings finally managed to get off the ground in a motorized flying machine, and Einstein opened the door to an unseen universe we are still trying to come to terms with. The scientific giants of the twentieth century extended the boundaries of human knowledge to a degree that dimmed the brilliance of all previous discoveries in human history. That dramatic expansion brought about a change in the way people regarded science. By the year 2000, the general public had begun to take scientific breakthroughs for granted and hardly blinked at some of the outrageous predictions for the near future offered up by self-proclaimed futurists beating their own drums.

There seems little doubt that the twenty-first century will indeed provide extraordinary advances in computer technology and biotechnology, although we should always keep in mind the so-called "law of unintended consequences." Pesticides, for example, which were supposed to be the answer to our planet's ever-growing need for more food, have ended up having almost catastrophic effects. We must also recognize that few things, including science, move forward at a steady pace in neat straight lines. Dead ends are everywhere, and quantum jumps are as common as step-by-step progress.

For all the wondrous developments of the twentieth century, a great many important mysteries remain unsolved. Some of those mysteries have tantalized the human race for hundreds, even thousands, of years. Aristotle, for example, was the first to give real thought to the migration of birds. He understood some things but got others completely wrong, and what he got wrong stifled further investigation for almost 2,000 years. We still have only partial answers to that mystery. In other cases, great breakthroughs of modern science have created problems of unprecedented scope and difficulty. The more we learn about the origins of the universe, for example, the more abstract the explanations become—to the point that many physicists have begun to regard them as closer to theology than science.

A hundred years ago we had no idea that the continents of the world not only move but have reshaped the face of the planet several times over—yet we still cannot accurately predict the earthquakes that those movements create. Eighty years ago no one was even asking how children acquire language, and although theories abound, we still don't know the answer. Just over 60 years ago, the possible existence of black holes was first suggested. We have now managed to confirm their existence, by inference, but in some ways their nature is more perplexing than ever.

We have failed to answer some ancient questions, and we have created deep new questions out of the need to find answers to long-standing ones. Sometimes it seems that the more we learn, the greater the validity of Hamlet's words on the battlements of Elsinore: "There are more things in heaven and earth, Horatio, than are dreamt of in your philosophy."

Chapter 1

How Did the Universe Begin?

Most major scientific theories come with the names of great scientific figures firmly attached to them. If someone says "gravity" the name of Sir Isaac Newton pops instantly to mind. "Evolution"? Charles Darwin. "Relativity"? Albert Einstein. But when the words "Big Bang" are spoken, no such name offers itself for convenient tagging. For the past few decades the Big Bang model has been widely accepted by cosmologists as the standard explanation of how the universe began, set forth in textbooks and general-interest magazines alike. Nonetheless, the concept is not associated with any one great scientist. At times it has been somewhat naughtily suggested by opponents of the theory that no one really wants to take credit for it. Indeed, the very term *Big Bang* was coined by one of its fiercest opponents, British astronomer Sir Fred Hoyle, as a way of lampooning the entire idea—but the name stuck anyway. In 1993, science author Timothy Ferris, astronomer Carl Sagan, and television reporter Hugh Downs were the judges for an international competition to come up with a better name for the theory. As Ferris notes in his 1997 book *The Whole Shebang*, nothing better was found among 13,099 entries from 41 countries.

The concept had its beginnings in a proposal by Georges Lemaitre, a Belgian monsignor of the Catholic Church, who became fascinated by physics and gained a Ph.D. from the Massachusetts Institute of Technology in 1927 at age 33. That same year, Lemaitre theorized that Einstein's laws of gravitation, spelled out in his 1915 general theory of relativity, implied that the universe must be expanding at the same rate everywhere and in all directions. Lemaitre further suggested that the universe had begun in the explosion of a primeval atom that contained all the matter in the universe. Edwin Hubble's subsequent discovery that distant galaxies were moving away from us and from one another in all directions, at speeds proportional to their distance from our own Milky Way galaxy, gave further credence to Lemaitre's theory. Hubble had not been aware of Lemaitre's concept, but the expansion of the universe, which he documented in 1929, nevertheless served to get more astronomers thinking about an initial explosion of some kind that could have created sufficient energy to create an expanding universe.

In the 1940s, physicists intrigued with the concept of an initial explosion theorized that immediately following such an event, the resulting plasma would have been far hotter than the interior of any star now existing, but it would have cooled over time, while still retaining at least a small amount of warmth. The residue of this process, they suggested, would create a thick haze that would still exist. This theory of what is now called the cosmic microwave background, or CMB, meant that the farther out in space (and back in time) we looked, the thicker the haze should be. This idea was largely ignored at the time because most astronomers and physicists didn't take the Big Bang theory very seriously, and in any case, there was no way to measure the CMB or confirm its existence.

In 1965, however, Arno Penzias and Robert Wilson of Bell Laboratories announced that they had detected a steady "hiss" of CMB radiation, which they had discovered by accident while developing a receiver for the first communications satellite, Telstar. That changed the thinking of a great many cosmologists. The Big Bang had been just another untestable theory before 1965, but now there was evidence of the kind of residue that such an initial explosion should have created. While many important scientists were converted to the theory of the Big Bang at this point, far more evidence was needed to back it up. Several predictions about the nature of a possible CMB had been made in the 1940s and 1950s. Investigations had calculated that it should have a temperature of about 3 degrees above absolute zero—the slight warmth that would remain after the cooling that allowed matter to coalesce out of the initial explosion. That warmth should also be *isotropic*—meaning, as Timothy Ferris has put it, "that any observer, anywhere in the universe, should measure the background as having the same temperature everywhere in the sky." Also, quantum physics appeared to demand that the CMB display a *black body spectrum,* emitting a maximum thermal radiation at a wavelength determined by its temperature—a spectrum that could be measured using specific quantum equations.

As the importance of the CMB became clear, the National Aeronautics and Space Administration (NASA) was persuaded to launch a microwave satellite designed to measure this "cosmic background." Free of the distortions of the Earth's atmosphere, the Cosmic Background Explorer (COBE) was expected to be able to peer back in time to the point about 500,000 years after Big Bang, when the universe cooled sufficiently to allow pure energy to start forming mass, thus making it possible for light to be released. Launched in 1989, COBE more than lived up to the hopes of cosmologists, providing evidence that the CMB was indeed isotropic, and that its temperature was close to 3 degrees above absolute zero (2.726° K). Moreover, it conformed to the expected black-body spectrum equations with astonishing precision.

By 1992, an all-sky map compiled by the COBE satellite had also substantiated another prediction: Matter, once it began to form from the cooling gases of the Big Bang, did so in clumps that would ultimately give birth to galaxies filled with individual stars. This was in line with the idea that microscopic quantum fluctuations in the early universe would disturb the generally homogeneous distribution of matter. In homespun terms, we are dealing with just slightly lumpy gravy—the flour is almost but not quite evenly distributed, and although the lumps may be few, they stand out.

Back in 1939, Hans Bethe, an American physicist, had shown that the heavy elements (in terms of their atomic weight) could be manufactured within the stars. These elements, of which the mass of planets and our bodies are composed, make up only 2% of the total mass of the universe. The rest is composed of about 75% hydrogen and 23% helium, with a trace of lithium. These light elements would have to have been created in the Big Bang, physicists calculated, in order to explain the abundance of hydrogen and the ratio of hydrogen to helium in the stars. The conversion of hydrogen into helium in the Sun alone releases 4 million tons of energy per second, and that process would create far more energy if the hydrogen/helium balance had not been established by the Big Bang itself. The heavier elements that were "smelted" in stellar furnaces would ultimately be thrown out into space, it was believed, to seed the universe with the raw materials of solid matter. The oldest stars, it followed, should retain less of the heavy elements because they would have been ejecting them for so long—which is just what was ultimately observed as new technology made such measurements possible. Thus, this distribution of elements, known as *cosmic element abundance*, also turned out to be right in line with Big Bang theory.

At this point, it might seem safe to conclude that the Big Bang theory had been proved correct. Whenever a new scientific theory makes predictions that can be tested, and those predictions are substantiated by observation or experiment, scientists rejoice in each succeeding confirmation. When enough such confirmations accumulate, the theory can be considered proved. But while the great majority of cosmologists accept the Big Bang, it is widely acknowledged that problems remain, which are serious enough in their implications to raise questions about the theory itself. Indeed, problems have cropped up so often that the theory has been in an almost constant state of crisis.

Fred Hoyle, who coined the term *Big Bang* with a derisive sneer, has always been a major opponent of the theory. In 1948, he had proposed what he called the "steady state" theory, along with Herman Bondi and Thomas Gold. According to this theory, the universe is immensely older than astronomical observations seemed to indicate, as it had always existed and always would.

Over vast eons of time, galaxies would be born, mature, and die, and new ones would constantly be born out of the resulting debris to take the place of the old ones. New galaxies would not necessarily form where the old ones had been, but the total mass of the universe would remain in balance. In this view, even the oldest galaxies we can observe are in fact quite new in terms of the larger picture. Many cosmologists disliked the steady-state theory because it suggested that we could never get to the bottom of things, and most physicists and astronomers are driven by the belief that we can. The fact that Hoyle could be abrasive in his comments, and was often described as arrogant by fellow scientists, didn't help matters. Nor did his great success with the general public as a popularizer. On the other hand, it can also be asked whether the belief that we can get to the bottom of things is not in itself the height of arrogance—certainly there seems enough of that characteristic to go around on all sides of these debates.

Hoyle's theory had its own problems, as well. For one thing, it made use of a modified form of the *cosmological constant,* a mathematical fudge factor Einstein had introduced into his theory of general relativity to reflect a universe that did not change. In 1929, Edwin Hubble, using his studies of the shift of color in distant galaxies toward the red end of the spectrum, called the "redshift," came to the conclusion that galaxies were flying apart at great rates with the expansion of the universe. Einstein's cosmological constant was no longer needed. Even Einstein called it the worst mistake he had ever made.

The antipathy toward the cosmological constant among most physicists, combined with the discovery of the CMB in 1965, appeared to put Hoyle's steady-state theory out of business. He wasn't about to close up shop, however. While there might be problems with his own theory, he continued to insist that there were even more problems with the Big Bang. Indeed, the Big Bang theory kept running up against new difficulties. One was that the more cosmologists learned, the clearer it became that the early universe did not work according to the laws of physics that now prevail. For at least the first 500,000 years after the Big Bang, until there was sufficient cooling to allow the formation of matter and the release of light (called "photo-decoupling" because light is carried by photons), the laws of our present universe did not exist. That discrepancy forced Big Bang theorists to turn to the notion that the initial universe was a *singularity,* a one-time event. Hoyle and his followers (for he had retained some) jumped all over this idea. Sure, they scoffed, you find something that messes up your Big Bang theory, and rather than doubt the theory you come up with a special exception that contradicts everything else we know.

Hoyle began to make some new headway of his own in 1990, when one of his followers, Halton Arp, an American cosmologist at the Max Planck Institute in Germany, pointed out that there have been a number of observations of redshifts that don't match up with their distance from the Earth. This was serious trouble. If the redshift was not after all a reliable indicator of the speed of the expansion of the universe, it would cut to the heart of Big Bang theory. Perhaps galaxies were not flying apart so fast, after all, and there would be no need for a Big Bang to set them in motion. Arp went further, saying in 1991, "It really gives the game away to realize how observations of these crucial objects have been banned from the telescope and how their discussion has been met with desperate attempts at suppression." Ignored evidence? Suppressed debate? The Big Bang theorists reacted with outrage. Meanwhile, as John Boslough notes in his 1992 book *Masters of Time,* several other physicists were charging that the Big Bang proponents were either ignoring evidence or developing hypotheses that couldn't be tested. Indeed, in 1986, Sheldon Glashow, who shared the 1979 Nobel Prize in Physics, joined with his Harvard colleague Paul Ginsparg to warn that physics in general was evolving into an activity so remote that it might end up being "conducted at schools of divinity by the future equivalents of medieval theologians."

The most significant of the untestable new ideas about the Big Bang was that of *inflation.* Proposed by Alan Guth in 1981, it held that at the very start, during what has been described as a "sliver of a second," the universe expanded at a rate exponentially greater than it now does, going from something analogous to a pinhead to the size of an orange or a softball in an infinitesimal amount of time. This may not sound like much, but mathematically it is staggering: The increase in volume was of a factor of 10 to the 50th power, or a 1 followed by 150 zeroes. After that instant of inflation, the universe settled down to the (relatively speaking) very leisurely rate of expansion that has since prevailed. In other words, at the very start the universe behaved like Superman for an instant and then decided to quit that stuff and amble around like Clark Kent for the rest of cosmic history.

To the general reader this can sound ludicrous, but the concept of inflation dispelled a number of dark clouds that were hanging over Big Bang theory, and it was widely welcomed. Among the problems it solved was that of the flatness of the universe. *Flatness,* as it is generally understood, is a somewhat unfortunate term to describe the physics involved in the theory, however much sense it may make mathematically. Physicists had determined that the universe ought to be either *open,* meaning that it would expand forever along an infinitely curved surface, or *closed,* meaning that eventually gravity would cause the universe to fall back into itself, presumably ending up in the kind of primordial atom that had given birth to the Big Bang. Unfortunately, however, there were no observable signs that it was either open or closed. It appeared to be perfectly balanced between these two possibilities, and that condition was described as flatness because the average curvature of space equaled zero, a "flat" trajectory.

To make things more complicated, the ratio of the universe's *actual density* (the amount of matter creating gravitational pull) to the density that would be required to cause it to collapse in upon itself, equaled one. The Greek letter omega was assigned to this ratio. Mathematically, an open universe would have a ratio that was less than omega, and a closed universe would have a ratio greater than omega. Thus, whether referring to curvature, with its value of zero, or the density ratio, with its value of one, the result was a flat universe. For the first time, Alan Guth's concept of inflation made that result reasonable. Never mind that inflation is often described in terms of a pinhead becoming an orange, which is assuredly round. Focus on the fact that the more a balloon is inflated, the more flat surface it has, and that because of the tiny instant of time in which inflation took place, it actually had a flattening effect. The mathematics, we are informed by Nobel Prize winners, work. (The mathematically challenged may simply prefer to think of an orange run over by a truck and let it go at that.)

Interestingly, one of the arguments against inflation accuses its proponents of "letting things go at that" on a cosmic scale. When Alan Guth was developing the concept, he ran into a problem that caused him to delay publication for two years. The theory predicted that such rapid expansion would have created a number of separate "bubbles." The walls of those bubbles should still be apparent, and they are not. In the end, Guth decided to publish anyway, in the hope that other cosmologists would be sufficiently interested to try to solve that problem. They were, around the world. Russian physicist Andrei Linde was the first to come up with an answer, which was subsequently also reached by others. He was able to show mathematically that the bubbles, which had been renamed "domains," would have developed independently of one another. What's more, our known universe would take up a mere billion-trillionth of just one of these "domains," and the walls of the bubble would be so distant as to remain forever beyond our observation. This calculation managed to remove an obtrusive elephant from the living room and to tether it conveniently out of sight behind the barn, but it was also the kind of thing that made Sheldon Glashow talk about medieval theology.

Nevertheless, like the idea of inflation itself, the bubble-domain theory was enthusiastically accepted among most cosmologists, including Stephen Hawking, widely regarded as the greatest living physicist. The bubble-domain theory, although untestable, solved problems of inflation (also untestable), and inflation had explained not only the flatness of the universe but also other difficulties with the Big Bang theory, including the fact that matter was so homogeneously distributed throughout the universe—the inflationary instant having acted as a kind of cosmic blender. To some critics, such as Halton Arp and Fred Hoyle, this is all far too convenient, however elegant the mathematics may be, however neat the dovetailing of theory with theory. But the critics remain rather lonely figures. Although many more individual physicists have difficulty accepting aspects of the Big Bang and the theory of inflation, they are willing to challenge the new orthodoxy only on smaller points, while being careful not to scoff at the whole.

For the moment, the Big Bang continues to reign as the best explanation for the origin of our universe. The emphasis should be on *our*. Don't forget those other domains, with walls forever beyond our ken. French physicist Trinh Xuan Thuan writes in his 1995 book *The Secret Melody,* "Our universe is just a tiny bubble, lost in the vastness of another bubble, a meta-universe, or super-universe, that is tens of million billion billion times larger. And that meta-universe is itself lost among a multitude of other meta-universes, all created during the inflationary era from infinitesi-mally small regions of space, all disconnected from one another." The grandness of this vision can be alluring or just mind-boggling. Some find it frightening. Others think it sounds like a religious concept, which can be reassuring or distressing, depending on one's beliefs. Some commentators have been at pains to point out that Georges Lemaitre, who had the first notion about what would ultimately turn into Big Bang theory, was a Catholic mon-signor first and a physicist second, whereas Fred Hoyle, champion of the steady-state theory, is an atheist. That may be too clever a distinction: It has also been said that some of Big Bang believer Stephen Hawking's work "eliminates the need for God."

As telescopes and computers become ever more powerful, capable of observing or simulating greater swaths of our universe, as quantum physics experiments delve ever deeper into the bizarre world of subatomic particles, it seems inevitable that the addi-tional knowledge gained will at times seem to support the Big Bang theory, while other discoveries confront it with new hurdles to overcome. In June 2000, a front-page story in the *New York Times* reported on a robotic telescope in Australia, which had pro-duced the first large-scale map of agglomerations of galaxies that form what can be thought of as cosmic continents. Enormous though these continents proved to be, their size did not exceed the predictions of Big Bang theory concerning such structures. The headline read, "Robotic Telescope Affirms Assumption On Universe's Birth." In the past, however, the *Times* has carried many headlines about discoveries that challenged other Big Bang

This photograph, taken April 1, 1995, by the Hubble Space Telescope, shows gaseous pillars in M16—the Eagle Nebula. The pillars are columns of cool interstellar hydrogen gas and dust, which act as incubators for new stars. They contain globules called EGGs (for "evaporating gaseous globules"), which are also more literally embryonic because they contain the embryos of stars, which will be uncovered through an erosion process created by the ultraviolet light emanating from massive newborn stars in the region. These spectacular columns are thus pillars of stellar creation. Courtesy NASA (Jeff Hester and Paul Scowen, Arizona State University).

assumptions. Some optimists, including Stephen Hawking, believe that we are close to understanding the whole of the universe, and that a "grand unified theory" is not far away. But even among champions of the Big Bang, there are many who suspect that we have only begun to understand how the universe works, and that we probably never will unravel its ultimate mysteries.

For now, the Big Bang is the standard theory. It is not yet truth.

✳ To investigate further

Ferris, Timothy. *The Whole Shebang*. New York: Simon & Schuster, 1997. Ferris is widely regarded as the best science writer in the business these days, and this book is a further feather in his cap. It is slightly more difficult to grasp than his earlier *Coming of Age in the Milky Way* but still very readable. Subtitled "A State-of-the-Universe(s) Report," it covers a host of cosmological issues but gives a particularly well-balanced account of Big Bang controversies.

Boslough, John. *Masters of Time*. Reading, MA: Addison-Wesley, 1992. Although new developments have changed the picture somewhat since this book's publication, it remains the clearest critique of the Big Bang theory available, detailing the crises the theory had faced in the 1980s, and pulling together the doubts of many eminent scientists, which are often parceled out in small doses and not picked up by the mass media. A science reporter with a distinguished career, Boslough emphasizes the continued validity of J. B. S. Haldane's statement of many years ago, "the universe is not only queerer than we suppose, but queerer than we *can* suppose." Like Ferris's book, Boslough's contains a very useful glossary of terms.

Thuan, Trinh Xuan. *The Secret Melody*. New York: Oxford University Press, 1995. A best-seller in France, where it was originally published (Thuan has also taught at American universities), this is an elegantly written book by an astronomer who fully accepts Big Bang theory and the concept of inflation. It is well-illustrated with charts, has several appendices that delve more deeply into the mathematics involved, and has a glossary.

Mitchell, William C. *The Cult of the Big Bang: Was There a Bang?* Carson City, NV: Cosmic Sense Books, 1995. This is an oddity, but an intriguing one. Self-published by an electrical engineer who worked on a number of NASA projects while with TRW, it is a flat-out attack on Big Bang theory. While the author has no credentials that would be accepted by most physicists, this book has hardly gone unnoticed. It has been endorsed by several cosmologists who themselves dispute the Big Bang, including Halton C. Arp of the Max Planck Institute, whose opposition to the theory is discussed in all the books listed here.

Note: Here, and throughout the book, sources are listed in the order of their usefulness in researching this book, with their potential as additional reading also taken into account.

Chapter 2

How Did Life on Earth Get Started?

I n the cosmic scheme of things, the Earth and the star it revolves around are Johnny-come-latelies. Our planet was formed out of the residue of the Sun's birth 4.6 billion years ago, whereas the universe as a whole is considered to have an age of 11 to 16 billion years. As is the case with the formation of all planets, the beginnings of the Earth were violent almost beyond imagination, and even after the globe itself took shape, the surface of our world remained molten for another 600 million years, superheated from within by its core and bombarded from without by asteroids that raised the temperature of the steaming oceans to the boiling point. Geologists call this period the Hadean interval of our planet's history, a time during which it was truly hell on Earth.

At some point after constant asteroid bombardment stopped, and the remaining asteroids settled into orbits that kept them mostly out of harm's way, various combinations of carbon, nitrogen, hydrogen, and oxygen were "reshuffled to produce amino acids and other basic biological building blocks." As Nobel laureate Christian de Duve explains in his 1995 book *Vital Dust,* "Brought down by rainfall, by comets and meteorites, the products of these chemical reshufflings progressively formed an organic blanket around the lifeless surface of our newly condensed planet." The resulting carbon-rich film was exposed to the effects of the continued churnings of the Earth itself, as well as to celestial objects that fell to the surface, and it was subject to ultraviolet radiation far greater than what reaches us today beneath our protective atmosphere. These materials were ultimately deposited in the seas, until, as brilliant British scientist J. B. S. Haldane wrote in a famous 1929 paper, "the primitive oceans reached the consistency of hot dilute soup." The main by-product of these processes was something sticky and brownish that has been termed "goo," "slime," and other names evocative of the childhood playground. Those who had long objected to Charles Darwin's original implication that we humans were related to chimpanzees and orangutans really went ape over this latest insult—we started off as slime!

So we've got soupy seas, and a lot of goo lying around everywhere. How did life arise out of these raw materials? Here is where the mystery begins. It is generally agreed that RNA—ribonucleic acid, a close relative of the DNA that determines our genetic heritage and that of all other living things—played a crucial role. Nonetheless, there are innumerable debates about the how, when, and where of life's actual start. Let's look briefly at some of the problems that have fueled such debates.

Biologists and chemists long believed that life would have taken at least a billion years to arise after the planet cooled down and the great rain of asteroids ended—about 3.8 billion years ago. This belief means that life on Earth is no older than 2.8 billion years, but increasing geological and even fossil evidence suggest that bacteria existed long before that. Greenland's Isua Formation, made up of the oldest rocks on Earth, dating to 3.2 billion years ago, contains carbon, the basic building block of all known life, in ratios characteristic of bacterial photosynthesis. Many biologists have come to accept that bacterial life must have existed even this early—and that if it did, then even more primitive organisms than bacteria must have existed still earlier. Bigir Rasmussen, a geologist at the University of Western Australia, has more recently found fossils of microscopic threadlike organisms that existed 3.5 billion years ago in Pilbara Craton in northwest Australia, as well as "probable" fossils that date to 3.235 billion years ago in volcanic vent deposits in western Australia. Such evidence carries with it a serious problem: The origins of life would then be pushed back to as few as 200,000 years after the end of the Hadean period, which seems to many biologists far too short a time for the chemical processes involved.

Rasmussen's more recent find, announced in *Nature* in June 1999, goes to the heart of another dilemma. Because the biomolecules basic to life, such as proteins and nucleic acids, are relatively fragile and survive longer at lower temperatures, many chemists have long insisted that life must have begun in a cold environment, even one that was below freezing. Yet Rasmussen unearthed the microscopic filaments in material that was originally close to a volcanic vent, meaning the temperature was extremely hot. Indeed, the most ancient organisms now alive are bacteria that live in still extant volcanic vents or springs where the water rises to a temperature of 230° F (110° C). The presence of these ancient vent bacteria strongly suggests the high-temperature environment favored by other scientists.

One of the cold-environment proponents is Stanley L. Miller, who achieved instant fame in 1953 when he carried out a series of experiments at the University of Chicago. He was then a graduate student, studying under the Nobel Prize–winning chemist Harold C. Urey. Urey had won the Nobel for the discovery of heavy hydrogen, also known as deuterium. Urey believed that the early atmosphere of the Earth was composed of a mixture of molecular hydrogen, methane, ammonia, and water vapor, and was particularly rich in hydrogen. (Notice the lack of oxygen except as a constituent of the water vapor: Life itself was necessary to produce oxygen in the atmosphere, through the emission of carbon dioxide during photosynthesis, thereby permitting the eventual development of more complex biological forms.) Miller created a sealed mixture of the elements Urey had proposed, and he bombarded it for several days with electrical discharges, simulating lightning. To his astonishment, a pinkish glow appeared in the glass container, and when he analyzed the results they contained two amino acids (components of all protein), as well as other organic substances thought to be produced only by living cells. This experiment, which his mentor had approved only reluctantly, not only made Miller famous but also created a new discipline, *abiotic chemistry,* focused on producing biological substances from conditions presumed to have existed before there was life.

The word "presumed" is crucial here. The presumptions about what Earth's atmosphere was like before life developed keep changing, and although a great many experiments have been carried out in the years since Miller's in 1953, nothing that can be called life has resulted, although important molecules of various kinds have been produced. As de Duve notes in *Vital Dust,* such experiments have often been carried out "under conditions somewhat more contrived than one would like for a truly abiotic process. In this rich crop, Miller's original experiment remains a paradigm, virtually the only one conceived exclusively with the aim of reproducing plausible prebiotic conditions, with no particular end product in mind." In other words, it is all too easy to adjust an experiment in ways that are more likely to produce *some* result, but the conditions themselves may be slightly too convenient. In any event, such experiments have not produced life, even in the most basic of forms—a single cell without a nucleus. As Nicholas Wade of the *New York Times* put it in his June 2000 article reporting Rasmussen's latest discovery, "The best efforts of chemists to reconstruct molecules typical of life in the laboratory have shown only that it is a problem of fiendish difficulty."

Major problems thus exist on two of the main research fronts that have been used to explore the puzzle of how life first developed. Not only is the date at which life first arose being pushed ever farther back, so far that it seems to allow too little time for the chemical changes necessary to create life, but those chemical reactions themselves remain as much of a mystery as ever. Indeed, despite extraordinary technical developments and a vastly increased knowledge of genetic materials, Stanley Miller's experiment of 1953 remains the cleanest example of such research. Even that breakthrough has been cast into doubt, in that many scientists now think that the balance of elements he used, based on the work of his mentor Harold Urey, was in fact incorrect. With changes in that balance, as tested in the laboratory, the production of the amino acids that he attained does not occur.

New difficulties have also clouded the picture of life's evolution that once seemed so clearly evident in the "family trees" of *phylogeny,* which traces the evolutionary history of an organism back to its roots. Evolutionary family trees, following the ideas of Darwin, were originally developed in the nineteenth century to show the history of groups of animals. The first complex family tree was drawn by German naturalist Ernst Haeckel, who also coined the term *ecology.* The discovery of DNA led to an ability to make such family trees not just of animals and plants, but also of the genetic materials of which they are composed, giving us a much deeper understanding of the processes of life. To create these trees, researchers use *comparative sequencing,* which involves determining the sequence of the molecular building blocks of nucleic acid (nucleotides) or of the amino acids in proteins, and then comparing the results with those obtained from other organisms. This technique has made it increasingly possible to discover the distance between two twigs on a family tree, in relation to the organism that gave rise to both, through the branching mechanisms of evolution or mutation. (This technique also helped researchers to determine the age of the still-extant ancient organisms now living in superhot volcanic vents.) The task of sequencing is perhaps most readily understood in terms of word puzzles in which a single long word is given and the player is asked to see how many shorter words can be put together from the available letters.

In the late 1970s, Carl Woese of the University of Illinois applied comparative sequencing to RNA molecules, which exist in all living things, to arrive at a more complex family tree than had previously been assumed. The resulting tree had clear branches delineating three fundamental kingdoms of living things: prokaryotes, archaea, and eukaryotes. *Prokaryotes* are microorganisms of the bacterial type. *Archaea,* the new classification proposed by Woese, is a second group of bacterial organisms generally found in very hot places such as scalding springs. *Eukaryotes* are organisms with large cells possessing a fenced-off nucleus, comprising all multicelled organisms such as plants and animals, including humans.

Since the early 1980s, however, as more genomes from all three kingdoms have been decoded, matters have gotten fuzzy. The pattern of trees based on genes other than Woese's original protein model are quite different. In addition, genes keep turning up that are surprising, even novel. This variation makes tracing all these genes back to common ancestors extremely complicated, and, even more troubling, suggests a primeval gene—the beginning of life—that is in itself quite complex, more so than a "starter" gene ought to be. The only plausible solution to this problem is to assume that instead of always branching upward, forming a vertical tree, some genes were exchanged horizontally during the early stages of the development of life. This idea is backed up by the fact that even now bacteria are capable of transmitting certain genes horizontally, including, unfortunately, those that make the bacteria resistant to antibiotics. That inference means that instead of having a nice straight trunk, the tree of life turns into something with a base resembling a Jackson Pollock painting. This is discouraging to say the least.

Undeterred, Carl Woese has suggested that the single-celled organism long believed to be the origin of life may instead have consisted of a kind of commune in which several kinds of cells exchanged genetic information horizontally in a rather sloppy way. That supposed sloppiness bothers some scientists. It means that it was only at some later point that cells developed the highly accurate replication of genes we see in DNA. The commune must have eventually turned into an upscale housing development in which each home has a different design—but when did that happen?

Experts are now assigning wildly different dates to the point at which the neat trees formed by DNA began their vertical branching—ranging from as recently as a billion years ago to the almost 4 billion years ago previously assumed. As with the Big Bang theory about the origin of the universe, theories about the origin of life on Earth have become more rather than less complex as new discoveries and modes of measurement have increased the level of knowledge. For that reason, other explanations for the origin of life on Earth that have long been disparaged as fanciful retain some adherents.

Could life have come to our planet from outer space? Certainly, asteroids, meteorites, and comets contain essential elements that form the building blocks of life, and there is general agreement that life on Earth arose from a combination of such materials—those already on the planet and those raining from space. But building blocks are one thing, life itself quite another. Some eminent scientists have put forward the idea that early life arrived here fully formed from outer space—not just the constituents of life, but the thing itself. As far back as 1821, Sales-Guyon de Montlivault suggested that the beginnings of life on Earth had arrived as seeds from the Moon, an idea that got renewed play in respect to Mars in 1890 when American astronomer Percival Lowell (who would accurately predict the existence of the planet Pluto) insisted that the red planet was crisscrossed by observable canals that could only have been built by intelligent beings. William Thomson (Lord Kelvin), who developed the Kelvin scale of temperature, proposed in the late 1800s that meteorites had brought life to Earth.

No one was as obsessed with such ideas as Svante August Arrhenius, the Swedish chemist who won the Nobel Prize in 1903 for work that established the basis for electrochemistry. His *theory of panspermia* held that bacterial spores could travel for vast distances through the cold of space in a hibernating state, ready to spring to life whenever they encountered a hospitable planet. He was not aware of the problem posed by deadly cosmic radiation. Fred Hoyle touted a variation on the idea of panspermia in connection with his steady-state theory of the universe, discussed in chapter 1. Hoyle went so far as to claim that such epidemics as the Spanish flu pandemic of 1918 were caused by bacteria from space, and that the human nose had evolved to help filter out such space-borne diseases. Francis Crick (who won the Nobel Prize in Physiology or Medicine in 1962, along with James Watson and Maurice Wilkins, for their discovery of the DNA double helix) joined with the prebiotic chemistry pioneer Leslie Orgel to go even further, backing the idea that life was "sowed" on Earth by an advanced alien civilization. They called this hypothesis "directed panspermia."

UFO enthusiasts are of course delighted to have a Nobel Prize winner such as Crick on their side—and science fiction writers have always pounced on such ideas. Lowell's Martian canals partly inspired H. G. Wells's famous novel *The War of the Worlds,* published in 1898. While many distinguished scientists hoot at the idea of panspermia, directed or otherwise, some are more cautious. Christian de Duve writes, "With such distinguished proponents, panspermia can hardly be dismissed without a hearing," although he goes on to note that no convincing evidence for such theories has turned up. This conclusion was reached in 1995, however, and the following year brought an announcement from NASA that caused headlines around the world.

The NASA announcement concerned one of a group of rocks that had been found in the Antarctic in 1984. The rocks were meteorite fragments, called SNCs (pronounced "snicks"), short for Shergotty-Nakhla-Chassigny, the locations where the first three such fragments were found. At the news conference, the single rock in question was displayed on a blue velvet cushion, and the head of NASA, Dan Goldin, started off by saying, "Today we are on the threshold of establishing whether life is unique to Earth," an excellent way to get the attention of journalists.

NASA scientists then told what was known for certain about this rock. Tests had established that it had been formed on Mars about 4.5 billion years ago. The rock had lain below the surface of Mars for half a billion years but had then been exposed to water after meteorites cracked the Martian surface. The rock had been subjected to a new experience some 16 million years ago, when the impact of an object from space, perhaps an asteroid, sent part of the Martian crust flying into space. After wandering in space for millions of years, that piece of crust had fallen to Earth in Antarctica a mere 16,000 years ago. Back in 1957, in a novel called *The Frozen Year,* science-fiction writer James Blish centered his story on a rock found in the Arctic, which turns out to be a remnant of a planet destroyed by Martians in a war of two worlds, causing the hero to exclaim, "Cosmic history in an ice cube!" The story that would unfold at the NASA news conference was somewhat less dramatic, although newspapers did their best to pump it up.

The NASA rock contained carbonates, similar to those formed on Earth by bacteria. Fine-grained iron sulfides and other minerals resembling bacterial products were found, too. Also, an electron-scanning microscope revealed tiny structures that could have

ALH84001,0

The meteorite fragment (called a SNC—"snick") pictured here was unveiled to the press by NASA in August of 1996. Found embedded in the ice of Antarctica in 1984, it underwent more than a decade of testing, which revealed that it had been part of the crust of the planet Mars, formed 4.5 million years ago, then pushed to the surface, and finally thrown into space about 16 million years ago by an asteroid. Within the rock were materials that appeared to be the fossilized remnants of Martian bacteria, suggesting that life was not unique to Earth. Courtesy NASA.

been fossils of Martian bacteria—they were too deeply embedded to have been formed on Earth. Not wanting to go out on a limb, NASA had a scientist on hand to say the structures were too microscopic to be bacteria, and that the carbonates had apparently been created at temperatures too high to permit life. His skepticism did nothing to prevent gigantic headlines screaming "LIFE ON MARS!"

Scientists have since debated the question in terms technical enough to frighten off any journalist. The matter could be proved one way or the other if one of those infinitesimal fossils could be sliced open. If evidence of a cell wall, or better still, cell division, existed, we would have an answer. Unfortunately, the techniques for carrying out such an investigation are not yet fully developed. When the answer does come, even if it is positive, there are certain to be many scientists who will still say it proves only that bacterial life once existed on Mars, as well as our own planet. It would not serve as proof that life had originated on Mars and traveled to Earth (or vice-versa), nor would it be evidence that the theory of panspermia was correct. Still, it would make it impossible to say any longer that there was no evidence of any kind to suggest such possibilities.

There may be more evidence, of an even more dramatic kind, about life in our solar system beyond the Earth forthcoming by 2015. The proposed NASA probe of Jupiter's moon Europa, which has a frozen surface suggesting the possibility that water exists at a great depth, could confirm that life is more common in the universe than conservative scientists suppose. We have learned in recent years that life exists on Earth at temperature extremes that were long thought inimical to biological organisms of any kind. If life of any kind were found in Europa's subsurface seas, it would elevate the concept of panspermia to a new level of seriousness. It would also further complicate the efforts of scientists to pin down the origins of life on our own planet, which are now stalled on two fronts: Theoretical approaches have been confused by increasing evidence that early life may have involved a lateral trading of genes, while laboratory experiments designed to create life out of chemical combinations have been frustrated at every turn. The state of the quest for an understanding of the beginnings of life on Earth are perhaps best summed up by the large headline in the *New York Times* section "Science Times" for June 13, 2000, which reported on the new fossils discovered in Australia: "Life's Origins Get Murkier and Messier."

⚛ To investigate further

de Duve, Christian. *Vital Dust*. New York: Basic Books, 1995. De Duve shared the 1974 Nobel Prize in Physiology or Medicine with Albert Claude and George Palade for discoveries relating to the structural and functional organization of the cell. He knows this material inside out and writes with great clarity. Add in a remarkable willingness to give a fair hearing to theories he does not agree with, and you have a book of great depth and scope.

Fortey, Richard. *Life*. New York: Knopf, 1998. Subtitled "A Natural History of the First Four Billion Years of Life on Earth," *Life* was a main selection of the Book-of-the-Month Club and, as might then be expected, is a book to be read rather than studied. Fortey is an important paleontologist, so there is plenty of science (angled toward his particular field), but he doesn't hesitate to spend a page discussing George Eliot's *Middlemarch* or Hollywood "blob" horror movies, always with pertinence. This is a delightful book from which the general reader can learn a great deal.

Margulis, Lynn, and Dorion Sagan. *Microcosmos*. Berkeley and Los Angeles: University of California Press, 1997. Although originally published in 1986, which means that some recent debates are not covered, its reissue in 1997 attests to the lasting strengths of this book, which received rapturous reviews when it was first published. Lewis Thomas, author of *Lives of a Cell*, wrote the introduction, in which he calls *Microcosmos* an "extraordinary" book for a general readership, and he is right.

Shopf, J. William. *The Cradle of Life*. Princeton, NJ: Princeton University Press, 2000. This book gets right into the thick of current debates on the subject, sometimes contentiously. Shopf is dubious about the processed carbon in the Isuan rock of Greenland, for example. This book is for readers with some scientific grounding, who want to explore the latest arguments.

Chapter 3

What Causes Mass Extinctions?

I t is estimated that some kind of life, even if only in the form of single-celled bacteria, has existed on our planet for about 3.5 billion years. That time span amounts to 20–25% of the probable age of the universe (see chapter 18 for the debate about *that* question). Impressive as that sounds, it took nearly 3 billion years after life's initial appearance for any real diversity to appear. As noted paleontologist David M. Raup succinctly puts it in his 1991 book *Extinction*, "All hell broke loose in organic evolution" about 600 million years ago. Since then, as few as 5 billion or as many as 50 billion different species have come into existence—the uncertainty about the number suggesting how little we know. While some of these species, such as the crablike trilobites, managed to hang around in one form or another for several hundred million years, 99.9% of all species that ever lived eventually died out. This is hardly the kind of survival record to brag about. What happened to the billions of species that are now extinct?

It is clear enough what happened to a number of species that have disappeared in very recent times: We human beings polished them off with great efficiency. To take just one example, there were millions of passenger pigeons in the United States during the nineteenth century—so many that flocks of them could darken the skies. To their detriment, however, they made good eating, and their feathers became popular for women's hats, so in 1914 the last of the species died in a zoo. More exotic creatures, such as the dodo, small in number to begin with, were hunted to extinction in the seventeenth century, and it is believed that the woolly mammoth met the same fate at the hands of primitive hunters before the last ice age. Currently, according to biologists and botanists, millions of species of animals and plants—the majority of them never even cataloged—are being extinguished as the destruction of rain forests in South America proceeds apace. All evidence suggests that of all the living creatures in the history of life on the planet, humans are the only ones that have had the capacity to drive numerous other species to extinction.

This awful fact accounts for only a tiny percentage of the extinctions that have taken place over the eons, however. We haven't been around to inflict such damage for very long, and billions of species disappeared without any help from us. The two major schools of thought about why extinctions are so common have been summed up in the full title of Raup's 1991 book *Extinction: Bad Genes or Bad Luck?*

During most of the more than 140 years since the publication of Charles Darwin's theories, the bad-gene school has been more prominent. Because Earth is constantly changing, its land masses moving almost imperceptibly to form new continents, its climate warming and cooling over fairly regular intervals, and even its very magnetic field undergoing complete reversals, the resulting earthquakes, volcanic eruptions, ice ages, and tropical heat waves would certainly present challenges to any living thing. Those with the genetic flexibility to adapt to such changes would naturally be expected to have the best chance of surviving, while those with the more inflexible genetic structure—which often meant the largest and most complex organisms—wouldn't make the cut. In addition, as any species moved toward greater genetic efficiency over long periods of time, its less well-adapted forebears would gradually die out, even without the added pressure of massive environmental changes. Even some quite primitive denizen of the early seas would have had an advantage over similar life forms if it developed the capacity to extract microscopic food more efficiently than the species from which it had evolved. Adaptability—the "survival of the fittest"—thus seemed sufficient to many scientists to explain most cases of extinction.

Over time, however, scientists have noted increasing problems with this approach, as more and more has been learned from the fossil record about the history of life on Earth. Evolutionary theory alone could not account for the five mass extinctions that have taken place, when the majority of life-forms existing at that particular time were wiped out. Thus, over the past half century, a growing number of scientists have been converted to the "bad luck" scenario, which holds that mass extinctions are brought about by rare catastrophic events of a nature and magnitude sufficient to devastate the entire planet. Before discussing the evidence for such catastrophic events, however, let's look briefly at the five mass extinctions that have taken place over the past 500 million years.

A mass extinction occurred in each of five different periods of geological time: the Ordovician, Devonian, Permian, Triassic, and Cretaceous (see chart). That leaves six other periods without a mass extinction, although there is no question that some extinctions took place throughout the entire 600 million year period, known as the Phanerozoic, during which complex life has existed on Earth. The only life that existed during the Ordovician period, which lasted from 505 million to 440 million years ago, inhabited the seas of the world. It was not until the Devonian period, from approximately 410 to 360 million years ago, that plants appeared on dry land, where they spread rapidly. Beginning with the Permian period, about 286 million years ago, vertebrates, both small and large, came into their own on land. There were both reptiles and mammals from the Permian onward, but mammals became much more diversified in the period following the extinction of the dinosaurs some 65 million years ago.

In his 1989 book *Wonderful Life,* and in other writings, Stephen Jay Gould has made it clear that the division of the Phanerozoic into separate categories such as the Age of Fishes, Age of Reptiles, and Age of Mammals is far too simplistic. Once both sea and land were occupied by biologically complex creatures, there was always

Geological Periods

Mass extinctions marked with an *
(mya = million years ago)

Phanerozoic Eon (544 mya to present)	Cenozoic Era (65 mya to now)	**Quarternary** (1.8 mya to present) Holocene (11,000 years ago to now) Pleistocene (1.8 mya to 11,000 years) **Tertiary** (65 to 1.8 mya) Pliocene (5 to 1.8 mya) Milocene (23 to 5 mya) Oligocene (38 to 23 mya) Eocene (54 to 38 mya) Paleocene (65 to 54 mya)
	Mesozoic Era (245 to 65 mya)	**Cretaceous** (146 to 65 mya)* **Jurassic** (208 to 146 mya) **Triassic** (245 to 208 mya)*
	Paleozoic Era (544 to 245 mya)	**Permian** (286 to 245 mya)* **Carboniferous** (360 to 286 mya) Pennsylvanian (325 to 286 mya) Mississippian (360 to 225 mya) **Devonian** (410 to 360 mya)* **Silurian** (440 to 410 mya) **Ordovician** (505 to 440 mya)* **Cambrian** (544 to 505 mya) Tommotian (530 to 527 mya)
Protozoic Eon	Precambrian Era (4,500 to 544 mya)	
Archean Eon	Beginnings of complex life approximately 600 million years ago	

a degree of overlap. Our fascination with dinosaurs may lead us to use phrases such as "when dinosaurs ruled the Earth," but in fact they did no such thing, despite the massive size some of them attained. There were only about 50 species of dinosaurs, whereas today there are 150 species of squirrels alone, and we don't suggest that squirrels rule the Earth even though they may sometimes seem to have our backyards annoyingly under their sway. Nor do we suggest that the largest of land mammals, the elephants, which are rapidly diminishing in number, rule the Earth. Size doesn't really count. Furthermore, if you were to go strictly by numbers, then insects rule the Earth and have since the Permian. It is far more accurate to say that diversity rules the Earth—a diversity that human beings constantly succeed in eroding despite the fact that our very lives depend on its continuance.

Although no one category of animal has dominated the Earth, during mass extinctions certain forms of life are always destroyed forever, as the dinosaurs were. It is also generally accepted that the mass extinction that finished off the dinosaurs, and vast numbers of other species, opened the way for mammals to grow in size and for one family of mammals to evolve into ourselves. Some scientists believe that if the dinosaurs had not been killed off, they might eventually have evolved into beings that walked upright, and could have finally developed an intelligence as great as—or greater than—our own. There is some evidence that smaller dinosaurs were already following a path that would have led to walking erect on two legs. Other experts demur, however, noting that the dinosaurs were around for a very long time without getting very far toward bipedalism, whereas the primates evolved very quickly, in relative terms, into human beings.

Leaving aside such speculations, the passing of the dinosaurs has proved to be the key to current arguments about what causes mass extinctions. There are two reasons for this: First, there is the great popular fascination with the dinosaurs, which has existed for the past century and a half, since the word was coined in 1842 by Richard Owen; second, because the dinosaurs disappeared in the last of the five mass extinctions, we have a better fossil record of their 140-million-year existence than we do of most forms of life wiped out in earlier periods.

Access to more information about an extinct species (or of a genus made up of a number of families of individual species) inevitably means that more scientists from a greater number of fields are likely to study it. Also, while ideas about dinosaurs changed considerably over the last decades of the twentieth century, and many mysteries remain (see chapter 6), these creatures carry a fascination that has hooked scientists with backgrounds that would hardly seem relevant to dinosaur study. No one has entered the dinosaur debates from further afield—and, partially for that reason, caused more of a ruckus—than Luis W. Alvarez, the Nobel Prize–winning physicist from the California Institute of Technology. Theories he developed, together with his son Walter, a geologist, shook up the field of dinosaur studies in the 1970s in ways that are still reverberating, and they opened an entirely new way of thinking about mass extinctions in general.

Back in 1973, Walter Alvarez and a group of other geologists were excavating in the area around Gubbio, in northern Italy, searching for evidence of reversals in Earth's magnetic field, which, for unknown reasons, occur about once every million years. At Gubbio, Walter Alvarez found a layer of clay almost devoid of fossils sandwiched between two layers of limestone with many fossil remains. It struck him as interesting that the clay layer coincided in geological time with the end of the Cretaceous period when the dinosaurs disappeared. (This period is often referred to as the K-T boundary, using the letter K to represent the German word for Cretaceous, *Kreide,* and the T as an abbreviation for Tertiary). In 1977, Walter returned to the United States and brought with him some samples of the clay layer, which he then discussed with his physicist father, Luis Alvarez.

Luis Alvarez had won the 1968 Nobel Prize in Physics for developing the liquid-hydrogen bubble chamber, which he used to identify numerous short-lived particles called "resonances." He was a man with numerous interests and achievements, however, having worked on the Manhattan Project to develop the atomic bomb and invented a radar guidance system for aircraft landings. The Gubbio clay samples intrigued him, and he began testing their geochemical makeup, got hold of some additional samples in 1978, and discovered that iridium in the clay was 30 times as concentrated as it was in the limestone layers above and below it. Iridium is a rare element on the surface of the Earth—but it is common in meteorites. The concentration of iridium in the clay from the end of the Cretaceous was startling.

Luis Alvarez considered several possible explanations. Perhaps, for example, a supernova in the near reaches of our galaxy at that period could have deposited iridium detritus on Earth—but the evidence did not support that hypothesis. Luis and Walter Alvarez turned to another idea: that a large meteorite had crashed into Earth. It would have had to be at least 6 miles (10 kilometers) in diameter to create dust clouds vast enough to blanket the Earth for several years, diminishing sunlight to the point that plant life, whether in the seas or on land, would have been widely affected. Had that happened, the resulting collapse of the food chain could certainly explain the demise of not only the dinosaurs but also the great numbers of other species that vanished at the same time.

The Alvarezes' theory was published in June 1980 in the journal *Science*. It was just the kind of dramatic scientific story that gets picked up by the popular press ("Meteorite Killed Dinosaurs!"), a development that only increased the annoyance of skeptics in various scientific fields. Many of Walter Alvarez's fellow geologists were particularly dismissive of the idea because they had developed their own theory involving massive volcanic eruptions—which could also create sun-blocking dust clouds. Other scientists thought the theory was plausible. It was also eminently testable. Could similar iridium deposits be found at various points around the globe, which would back up the findings from Gubbio? Was there a crater large enough, and of the right age, to prove that such a meteorite had hit the Earth?

Within two years, the presence of iridium in strata of the correct age at other widely disparate locations had been established, but other scientists began to ask new questions. One study raised serious doubts that iridium could remain in the atmosphere long enough to be carried around the globe from a single impact site. Computer models, however, showed that a "ballistic dispersal" of iridium was feasible. As these debates continued, there remained a greater problem: Where was the requisite crater to be found? None of the great land craters were the right age or size. Then in 1989, an undersea crater was found at the north coast of the Yucatán Peninsula by oceanographers mapping the region. Measurements

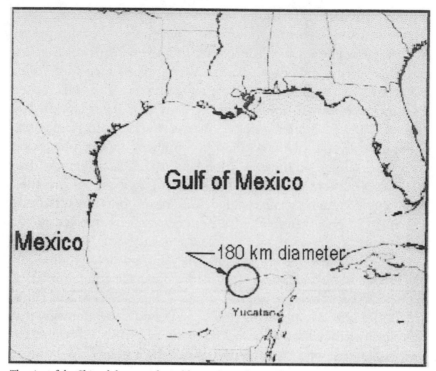

The size of the Chicxulub crater, formed by an asteroid 65 million years ago, was long underestimated because nearly half of it lay beneath the sea, and most traces of it on land had been eradicated by erosion and changes in the shape of the Yucatán Peninsula over the eons. Once its full extent (112 miles [180 kilometers] in diameter) and age had been determined, it backed up the theory that the extinction of the dinosaurs had been caused by a catastrophic asteroid impact. Courtesy U.S. Geological Survey.

of this Chicxulub Crater were carried out, and in 1993, it was announced that it had a diameter of 112 miles (180 kilometers), larger than the state of West Virginia, and indeed the largest known crater in the solar system. What's more, tests showed that it had been formed 65 million years ago, right on the button in terms of the mass extinction that eliminated the dinosaurs. For another four years, material brought up from the crater was tested, and in 1997, other researchers concluded that the deposits of iridium and other elements were consistent with the geological findings at Gubbio, Denmark, and New Zealand, which Luis and Walter Alvarez and their colleagues, Frank Asaro and Helen Michel, both chemists, had announced in 1980. At that point, most scientists granted that a meteor impact had played at least some role in the extinction of the dinosaurs. This conclusion was buttressed by studies announced in November 1996, indicating that the Yucatán meteor crashed into the Earth at a sharp angle that would have created a huge firestorm over North America.

That conclusion did not mean that the larger debate about mass extinctions was settled, however. Some scientists, including David Raup, concluded that the physical mechanism to explain all mass extinctions was now clear. Scientists would not have expected to find craters that had been created at a time coinciding with earlier mass extinctions: The changes in the Earth's surface over millions of years would inevitably have erased all traces of impacts that had taken place at the time of earlier extinctions. In the course of the twentieth century, the discovery of continental drift had led to evidence showing not only that there had been a single supercontinent, called Pangaea, 200 million years ago, but that even that supercontinent had been formed from the pieces of a still earlier supercontinent, called Rodinia. These epic reorderings of the Earth's surface could certainly provide ample reason for the lack of other well-documented craters like Chicxulub.

Despite these findings and inferences, the attempt to suggest that all five mass extinctions over the past 500 million years had been caused by meteor impacts was provocative. Scientists who had remained at least partially skeptical about this scenario, even in regard to the dinosaurs, were prompted to speak up more loudly. These scientists were willing to admit that the impact of a meteor had played a part in the extinction of the dinosaurs, but only a part. The plateau region of western India is marked by vast lava deposits, known as the Deccan Traps, and a number of scientists argue that such increased volcanic activity could have produced atmospheric conditions as inimical as the impact of any meteor. While the dating of the Deccan Traps is somewhat problematical, some suggest that a combination of both volcanic eruption and meteor impact might have been necessary to tip the scales. Taking another tack, other experts have insisted that the dinosaurs died out more slowly in other parts of the world than they did in North America, where the impact of the Yucatán meteor would have been greatest. Still another point of view suggests that the dinosaurs were already beginning to disappear before the meteor struck, and that they would have vanished eventually without such a catastrophic event to hasten their departure. This last point of view is often linked to the idea that many of the dinosaurs had simply gotten too big for their own good, and that only modest changes in the environment would have been sufficient to create a scarcity of food. The smaller dinosaurs, by this way of thinking, were already evolving into animals like modern reptiles, as well as into the first true birds.

The argument that some dinosaurs had grown to excessive size while others were evolving into new species supports the bad-gene theory of extinction. Gigantic size could have been a bad genetic trait because it created increased vulnerability to environmental changes, while smaller species might have been better able to adapt over time. Even David Raup concedes that some species have always died out because of species-specific genetic problems. Such problems could range from diseases that affect only one species, or a very few species, to habitat changes that proved lethal to species that had filled a very narrow niche. Both these problems have been evident in our own time in regard to various endangered species, such as the snail-darter (habitat destruction) or the Florida panther (hereditary deformity of the sex organs). Raup himself believes that the trilobites were affected by bad genes. Six thousand species of trilobites have been found in fossils of the Cambrian period, with those numbers decreasing sharply during two subsequent mass extinctions, and disappearing altogether at the end of the Paleozoic era 325 million years later.

Raup argues persuasively, however, that bad genes cannot account for the vast numbers of species wiped out during mass extinctions. Something has to have happened in these situations that kills off species with good genes, as well as bad. Raup himself takes responsibility for the much quoted figure of 96% species extinction at the end of the Permian, which was derived from an article he published in 1979, in which he put forward that number as an upper limit, with many caveats attached. Still, even a 70% extinction rate is more than massive enough to suggest a cataclysmic event.

Nonetheless, Raup's belief that meteor impacts were the chief cause of all five mass extinctions does not sit well with many scientists. He does have his supporters, though, and it is possible to counter some objections to his theory. To those who insist that greatly increased volcanic activity played a recurring major role (and who do have geological evidence to support their view in some cases), it can be suggested that a large enough meteor impact might well create volcanic activity in itself, turning it into an effect rather than a cause. Even so, some experts are happiest with the idea that any given mass extinction can have several overlapping causes. Others think that a single major cause was at the root of each of the five major extinctions, but that that cause probably differed for each one. In one case, it might have been extreme volcanic activity, in another a drastic rise in the sea level, and a third might involve severe climate disturbances. One of these scenarios, including a meteor impact, might have occurred more than once.

It is unlikely that these debates will ever be settled. While the Yucatán crater serves to give great credence to the meteorite theory in regard to the most recent mass extinction, researchers hold little hope for finding such evidence for earlier mass extinctions. The surface of the Earth has simply changed too drastically and too many times over the past several hundred million years. No doubt, discoveries of other kinds will tilt the debate in one direction or another in the future, at least for a time, but ultimate answers seem elusive.

Some experts suggest that we might come up with further answers the hard way. The greatest extinction since the one at the end of the Cretaceous, 65 million years ago, is now underway. We humans are causing it, and some scientists worry that we may be creating an environmental collapse that will bring on our own extinction—an object lesson we could clearly do without. On another score, we could be confronted with a replay of the Cretaceous catastrophe if a large enough meteor hit the Earth again. Wandering asteroids are out there, and we know about some near misses. Few astronomers doubt that Earth will suffer a massive hit again, sooner or later. Unless we develop a program to break up such an asteroid in space, perhaps using atomic bombs, as a number of scientists suggest can be done, we could find out firsthand what kind of changed world the dinosaurs suddenly faced. Leaving aside such dismal ways of attaining knowledge about how mass extinctions occur, the first four mass extinctions will remain mysterious, their causes endlessly debated, with some certainty established only in regard to the fifth and most recent catastrophe.

✳ To investigate further

Raup, David M. *Extinction: Bad Genes or Bad Luck?* New York: Norton, 1991. Raup has been described by his colleague and friend Stephen Jay Gould as the "primary architect" of the increased scientific interest in extinction over the past few decades. This book, written for a popular audience but including many sophisticated charts, is lively, provocative, and thorough.

Alvarez, Walter. *T. Rex and the Crater of Doom.* Princeton, NJ: Princeton University Press, 1997. This scientific detective story gives the complete background on the meteor impact theory developed by Walter Alvarez and his father, the late Luis W. Alvarez. It is a fascinating tale of scientific discovery and what it takes to convert doubters to the validity of a new theory.

Gould, Stephen Jay. *Wonderful Life.* New York: Norton, 1989. A best-seller that is regarded by many as a classic work, it uses the fossils of the Burgess Shale in British Columbia as the basis for a wide-ranging analysis of the meaning of both evolution and extinction. There is currently a backlash against Gould's views, led by Richard Wright, author of *Nonzero: The Logic of Human Destiny* (New York: Pantheon, 2000), on the grounds that Gould overemphasizes the "accidental" aspects of evolution, including that of human beings. It is a debate worth paying attention to, and *Wonderful Life* is the place to start.

Wilson, E. O., Ed. *Biodiversity.* Washington, DC: National Academy Press, 1988. This famous collection of research and essays reports on ongoing and possible future extinctions. This is an academic book that those with a real interest in the question of whether we are creating a new kind of mass extinction right now will find important.

Chapman, C. R., and D. Morrison. *Cosmic Catastrophes.* New York: Plenum Press, 1989. A well-written book that discusses, in scientific detail, the risks of Earth being hit by a massive meteor in the near future.

Wade, Nicolas, Ed. *The Science Times Book of Fossils and Evolution.* New York: Lyons Press, 1998. This collection of articles originally published in the *New York Times* offers clear, solid accounts of numerous developments in the fields of fossil analysis and evolutionary theory during the 1990s.

Chapter 4

What Is the Inside of the Earth Like?

April 17, 1906, had been another triumphant night for the great operatic tenor Enrico Caruso. He had been endlessly cheered following his performance at the opera house in San Francisco, a city that even then had a large Italian contingent. Whenever he had visited, he had felt very much at home there, but by the next morning he was vowing never again to return to the entire state of California, let alone San Francisco. At 5:13 A.M., a massive earthquake struck, and Caruso barely escaped alive from his collapsing hotel. For three days following the devastation of the quake itself, the city burned. An outbreak of bubonic plague was brought on by rats whose nests had been destroyed along with most of the city's buildings. Then there were the stories about the cats of San Francisco, which began to appear in the press. Many people were reporting that their cats had gone berserk just before the quake struck. Had they known what was going to happen before the earth shifted enough for human beings to notice? Some people were impressed enough to buy a cat as a kind of warning-signal for the next possible earthquake.

Scientists were interviewed on the subject of the pre-earthquake behavior of the cats. "Nonsense," said the scientists. "Old wives' tales." "Hysteria." And the idea that cats might be able to sense

A view of San Francisco taken at 10:00 A.M. on April 18, 1906, five hours after the earthquake. The great opera tenor Enrico Caruso barely escaped from the collapsing Palace Hotel, lower left-hand corner, during the quake. Despite great advances made during the past century concerning the inside of the Earth and the movements of tectonic plates, earthquake prediction still remains almost impossible in terms of meaningful time frames. Courtesy NOAA-EDS.

the coming of an earthquake was relegated to the realm of folk-lore. San Francisco, rebuilt with stringent new construction codes, would survive many secondary earthquakes over the subsequent decades. Then on the evening of the second World Series game of fall 1989 between two area teams, the San Francisco Giants and the Oakland Athletics, another massive earthquake stunned the city. A nation tuned in for the televised game held its breath as buildings rocked, highways collapsed, and fires broke out. No earthquake in human history had been viewed by so many millions of people as it was happening. In the aftermath, the old stories about cats came back. The 1906 earthquake had happened in the middle of the night, and it had been easy to dismiss the cat stories then as the tales of unreliable types, half of them probably inebriated. This time, however, the reports of cats racing around in a greatly agitated state just before the earthquake came from off-duty police and fire-service officers, sitting at home awaiting the start of the game. Laboratory technicians, doctors, and other highly trained people told similar stories. The cats of San Francisco had once again gone crazy. This time, scientists paid attention. Maybe there was something to these old wives' tales after all, so research programs were started to track the behavior of cats during earthquakes.

The fact that it was deemed wise to look into earthquake-related feline behavior at the end of the twentieth century tells us something about the state of earthquake prediction: It is virtually nonexistent. Yes, seismologists can confidently predict that, for example, a catastrophic earthquake is due in the Los Angeles area sometime soon—but "soon" means maybe tomorrow and maybe 30 years from now. Earthquake forecasting makes weather forecasting look extremely accurate, despite the cool front that doesn't arrive or the snow that appears as if from nowhere. Even so, we now know vastly more now than we did at the time of the San Francisco earthquake of 1906.

It was not until 1912, for example, that the concept of continental drift was first proposed by Alfred Wegener, a German scientist born in 1880. Previously, everyone had assumed, scientist and nonscientist alike, that the continents had always existed as they now do, from the time Earth took permanent form. Wegener became interested in *meteorology,* the then-new science of atmospheric studies, and went on explorations of Greenland. Supposedly, the floating ice in the waters around Greenland gave him the idea that land masses might also move around the Earth. He started looking for evidence to support this theory, and found it in two kinds of connections among different continents. One connection was geological: deposits of the same age and kind in places separated by wide oceans. Second, similar ancient fossilized animals and plants appeared on different continents, even though in the twentieth-century world such similarities seldom existed, with each continent having flora and fauna specific to it. Tomatoes, corn, and potatoes were indigenous to the New World of the Americas, while cabbages, eggplants, and zucchini were indigenous to Europe, and the same kind of exclusivity could be seen in the case of many animals. In the very distant past, however, certain plants and animals had existed on more than one continent. A prime example was the Glossopteris seed fern, which lived 270 million years ago on the present-day continents of South America, Africa, Australia, and Asia. It was clear to Wegener that this must mean there had once been a single supercontinent, and he published a book called *The Origin of Continents and Oceans* in

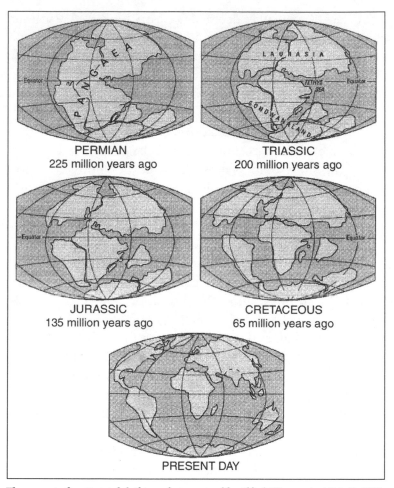

The concept of continental drift was first proposed by Alfred Wegener in 1912. In 1915, he published a book proposing that there had once been a single massive continent, which he called Pangaea. Originally dismissed by most geologists, Wegener's ideas were ultimately confirmed, both by geological sampling and by the discovery of tectonic plates that provided the mechanism for continental drift. This series of maps shows the changing surface of Earth over the eons. Courtesy U.S. Geological Survey.

1915, which laid out his theory in detail. All the continents on Earth had once been a single mass that he called Pangaea.

Some scientists were fascinated by this idea, and they found Wegener's evidence quite compelling. This relatively small group came to be known as the "mobilists." But most geophysicists followed the lead of their eminent British colleague Sir Harold

Jeffreys, who found the whole notion preposterous. His own observations of earthquakes had convinced him that the interior of the Earth was absolutely rigid. Moving continents indeed! Unfortunately, the mobilists were unable to provide any plausible mechanism that would make it possible for continents to move.

It was not until the 1960s, three decades after Wegener's death, that such a mechanism was discovered. This discovery would probably have taken much longer, as Simon Lamb and David Sington point out in their 1998 book *Earth Story,* had it not been for the development of nuclear submarines in the 1950s. Given this development, it became imperative for the first time to have maps not just of the surface of the seas but also of the ocean floor. The U.S. Navy provided the ample funding necessary for this massive project, which made extensive use of new echo-sounding technology to record the vibrations that occurred when small explosives were dropped into the water. It had been assumed that the ocean floor, worn down over millions of years by the motion of the water and the polishing effect of sediment, would be quite smooth. That turned out to be completely wrong.

The most astonishing discovery was that there was "a virtually continuous undersea mountain range which snakes right round the planet," as Lamb and Sington write. "In fact, this is the longest mountain range on earth." What's more, the ocean floor had fracture zones as though it were splitting apart in enormous slabs at right angles to the undersea mountains. The implications were enormous. The ocean floor was clearly much less ancient than had been believed, and it was subject to the same kind of volcanic and earthquake activity as the continents themselves. In 1960, Harry Hess of Princeton University began putting together these new discoveries with earlier ones that had long intrigued him, creating a new theory of what was happening to the surface of the Earth below the oceans. The midocean ridges were obviously rising, but there were also flat-topped underwater "islands," which he called "guyots." These were apparently sinking, but their tops might once have protruded above the ocean, where they could have become flat because of erosion. That had to mean that the rocks on the ocean floor were denser than those on land, causing them to sink into the upper mantle between the

Earth's surface crust and its core—with the upper and lower mantle consisting of minerals that increase in density at the lower level. The deep ocean mountains, Hess believed, must have been pushed upward because of some interior force, then slid to one side before starting to sink again. He saw this as a giant conveyor belt that continually changed the form of the ocean floor. The ocean floor was not flat and not stationary, but continually remaking itself.

Fred Vine, a graduate student at Cambridge University in England, heard a famous lecture given by Hess and developed Hess's ideas even further. Vine had been assigned to analyze the results of a British magnetic survey of the Indian Ocean. This field of inquiry had recently led to conclusions that the Earth's magnetic field had reversed several times in the history of the planet. A compass now points to the North Pole, but when the magnetism was reversed, it would have pointed to the South Pole. Sir Edward Bullard had suggested that the outer core of the Earth, between the mantle and the solid inner core, must consist of liquid (molten) iron, and that the flow patterns of this molten layer would create a *dynamo action,* which could indeed reverse itself at various times. The proof of that idea came with a new technique for dating rocks based on the analysis of the radioactive decay of argon gas trapped in volcanic lava as it cools into rock. Carried out at the University of California at Berkeley, these difficult tests showed that the magnetic field of the Earth did indeed reverse itself approximately every million years.

In 1963, Vine, together with his supervisor, Drummond Mathews, came to the conclusion that there must be a double conveyor belt on the ocean floor, with one on either side of the midocean mountain ranges, gliding apart and creating a striped effect each time the magnetic field of the Earth reversed. Further work by such scientists as J. Tuzo Wilson and Alan Cox backed up the ideas of Hess and Vine, making clear that the adjacent parts of the ocean floor were constantly sliding past one another along the fracture boundaries. What's more, none of the rocks composing the ocean floor proved to be more than 200 million years old, nearly 10 times younger than land masses.

All this slipping and sliding of the ocean floor led many scientists to begin to take a fresh look at the long-ignored ideas of

Alfred Wegener concerning his single continent Pangaea. If this much movement was taking place undersea, might not the continents themselves have moved, even if at a much slower pace? Numerous geologists and geophysicists began turning up more and more clues to what was really going on. An earthquake near Anchorage, Alaska, in March 1964, measuring a massive 8.6 on the Richter scale, occurred in an area under the jurisdiction of geologist George Plafker of the U.S. Geological Survey. Studying the aftermath of the quake, he became convinced that because there was no observable fault line of sufficient size on land in the affected area, it must lie in the sea off the coast. Over the next few years, Plafker and other geologists were able to show that in very specific areas around the world, the ocean crust was sinking lower into the Earth's interior, even as it slid forward, and at some point it would slip under the mainland crust, pushing it upward and causing an earthquake.

Thus was the theory of plate tectonics born. The plates, some large, some secondary, form what is called the *lithosphere,* the outer shell of the Earth. The crust on which we farm and build cities is just the top of the lithosphere, about 220 miles (352 kilometers) thick, on average. These lithospheric plates move, and that movement has since been measured, thanks to satellite technology. The movement is very small, generally less than half an inch per year, but that adds up over the millennia, and at various points, the passage of one plate beneath or over another (or sometimes past one another) creates a sudden lurch that results in an earthquake. The pressures that the two plates are exerting on one another suddenly become too much. Something has to give—and a larger movement of the planet's crust occurs as one plate or the other ruptures.

Between the San Francisco earthquakes of 1906 and 1989, geophysicists finally came to understand much more completely why these ruptures of the crust of the Earth occur. They also know, around the world, where the greatest danger spots lie, because it is clear that two tectonic plates are grinding against one another in those places, building up intolerable pressures. California's San Andreas Fault gets the most attention from the media, at least in the United States, and comedians can always get a laugh about "Southern California falling into the sea." But the laugh is a nervous one because there is no doubt that a massive earthquake is due—sometime.

Solving the mystery of tectonic plates has helped to make other aspects of the Earth's composition clearer as well. But the further below the surface we get, the more speculative the science becomes. The outer crust, or shell, of the planet is as much as 200 miles (320 kilometers) thick on the continents but extends only about 15 miles (24 kilometers) below the ocean floor. The crust rests on an upper mantle of minerals known as olivine and peroxide, with some garnet; the lower mantle consists of similar rocks of even greater density because of the pressure brought to bear on them. This pressure, and the accompanying rise in heat at greater depths, is sufficient to turn carbon into diamonds. Diamonds that are mined have been ejected from the lower mantle in volcanic eruptions, embedded in the molten lava that becomes basalt when it cools. Some diamonds are believed to have been ejected from as far down as 600 miles (960 kilometers) below the surface of the earth. It is only by using diamonds, put under great pressure and heated with laser beams in laboratory experiments, that scientists have been able to create minuscule quantities of the incredibly dense mineral structure called perovskite, which makes up the lower mantle.

Below the mantle is an outer core of liquid (molten) iron and nickel, sloshing around at temperatures close to those of the interior of the Sun—or so it is believed. There are good reasons to believe this must be the case, but at such depths within the planet, inference is all that scientists have to go on. Vibrations called P waves and somewhat slower S waves travel through the Earth during earthquakes, and these can be measured by seismographs. These waves have certain qualities that reveal the kind of material they are moving through. The molten outer core of the planet surrounds a solid inner core of nickel and iron. Why doesn't the molten outer core melt the inner core, too? The assumption is that at some point the inferno-like temperatures of the outer core drop off significantly. This is apparently due to a convection pattern (which can be reproduced in the laboratory at far lower temperatures) that causes the hottest material to move upward in plumes, displacing cooler material at the top, which then sinks to the bottom.

The Earth is a living planet not just in terms of its surface ecosystem of animals and plants, which thrive in a propitiously balanced atmosphere dependent on the fact that the majority of the surface is covered by water. In contrast, the Moon is a dead world, both inside and out, although it is thought that it, too, had a molten core for millions of years after it was (apparently) split off from the Earth due to a collision between our world and another smaller one closer in size to Mars. Mars is a dying planet. It has become clear that it once had large seas, and enough water still remains to create a thin atmosphere and ice at its poles. It may have a great deal more water trapped beneath the surface as permafrost. We do not know what happened on Mars to cause it to begin to die—if we did, we might be far more careful about how we treat our own planet. On the other hand, it may also be that Earth alone in our solar system had the right combination of conditions to succeed as a terrestrial planet—a world with a hard surface that was able to retain its water. We shouldn't be parochial about this view, however. The gas giant, Jupiter, apparently lacking in any kind of surface crust, is also alive in its own way, as the Great Red Spot and other vast storm systems attest. Still, the Earth is clearly unique in this little corner of the universe.

That very uniqueness, however, is precariously balanced. It is believed that the liquid outer core is in itself inherently unstable in some ways—that instability would account for the flipping of the Earth's magnetic field every million years or so. Also, the fact that the Earth is alive means that it changes constantly. Alfred Wegener was right about Pangaea, the supercontinent. By the 1980s, geological evidence had proven conclusively that South America and Africa were once part of the same land mass, as were all the other continents. Indeed, any child can look at a flat map of the world and see that Africa and South America fit together very nicely, thank you, like the pieces of a jigsaw puzzle. People had noticed that before Wegener's time, of course, but they put it down to coincidence or God's will. It has now been quite well established that before Pangaea existed, there were separate continents, differing in shape from our current ones, and that before that, there was another supercontinent, which has been named *Rodinia,* after the Russian word for "motherland." Some geological research even suggests that the entire process happened at least once before *that.*

A planet that is able to reshape entire continents several times over—even if it does take a billion years or so—is obviously alive in ways that stretch the human imagination to its utmost. The history of complex life on Earth goes back only about 600 million years. Long before that, however, the entire planet was busy forming and reforming itself into different patterns. We just happened along at a point that gave us geological niceties such as the Straits of Gibraltar and the white cliffs of Dover. Only 30,000 years ago, the Bering Strait, between Russia and Alaska, was not water but land—that's how the people whom we know as Eskimos (or more properly, Aleuts) and American Indians (Native Americans in terms of recorded history, but themselves travelers from another continent) got here in the first place.

Against this epic backdrop, the fact that we have even figured out how earthquakes happen is quite remarkable. Will we ever be able to pinpoint when earthquakes will shatter our cities, or when volcanoes will start building new mountain ranges regardless of the towns that lie at their bases? When the whole history of the living planet Earth is taken into account, it may sometimes seem a form of hubris even to try. Maybe the cats of San Francisco know.

✳ To investigate further

Lamb, Simon, and David Sington. *Earth Story.* Princeton, NJ: Princeton University Press, 1998. Based on a BBC television series, this book offers a comprehensive look at "The Shaping of Our World," as the subtitle puts it. Profusely illustrated in color and very clearly written, it manages to include considerable detail for a book aimed at a popular audience.

Zebrowski, Ernest J., and Ernest Zebrowski Jr. *Perils of a Restless Planet: Perspectives on Natural Disasters.* New York: Cambridge University Press, 1999. A wide-ranging inquiry into natural disasters of all kinds, this serious but entertaining book examines the scientific efforts to understand such events, as well as their consequences for society.

Harris, Stephen L. *Agents of Chaos: Earthquakes, Volcanoes, and Other Natural Disasters*. Portland, OR: Mountain Press, 1990. This popularly oriented book on natural disasters in the United States provides good material on earthquakes and the difficulty in predicting them.

Menard, H. William. *Ocean of Truth—A Personal History of Global Tectonics*. Princeton, NJ: Princeton University Press, 1995. Bill Menard was one of the pioneers of plate-tectonic theory, and his various sea voyages with scientific teams, starting in the 1950s, make compelling reading. This book is particularly good for those who like to know the background story about the ways in which scientific breakthroughs are achieved.

Chapter **5**

What Causes Ice Ages?

We live in what is called an *interglacial period,* one of the warmish valleys of time between the more usual icy peaks of the past 35 million years. What are known as the *temperate regions* of the Earth get snow in winter, which disappears in the spring. At present, vast ice-covered regions cover both poles, which means that in terms of Earth's climatic history, our present era is on the cool side. The Earth was far hotter than it is now during the 250 million years when the dinosaurs roamed its surface, for example, when trees grew near the North Pole. In terms of the more recent past, the last ice age ended about 12,000 years ago. Some 20,000 years ago, it was so warm that hippopotamuses were wandering around Hertfordshire in southeastern England. The discovery of the bones of these jungle animals in the nineteenth century was one of several developments that led scientists to begin to realize how different the climate of the Earth had been in the past, sometimes approaching the tropical in what are now cool northern latitudes, and sometimes so cold that ice sheets covered much of North America, as far south as New York and Illinois.

Quite aside from the matter of hippos in Hertfordshire in the 1800s, geologists finally began to take note of the confusing jumble of out-of-place rocks, fossils, and peculiar petrified oyster shells that could be found throughout England, northern Europe, and the upper parts of North America. Where had these varied materials come from? One famous British geologist, William

Buckland, thought they must have been deposited by the biblical flood, but others would soon develop more scientific ideas. The first to make the connection between glaciers and the jumble that had come to be known as "drift" was Jean Louis Agassiz. This Swiss scientist, who started out as a zoologist and subsequently laid many of the foundations of modern geology, influenced a generation of scientists as a professor at Harvard, following his emigration to America in 1846. While exploring a glacier in the Swiss Alps in the late 1830s, he noted that it had shrunk in recent years, leaving a recent deposit of the kind of ancient residue that could be found all over Europe. That led him to conclude that glaciers must have once covered a far greater area than the Alps or the northern areas where they existed in the nineteenth century. As later geologists started digging down, they further realized that there had been many such layers of drift. That meant that glaciers must have moved down over Europe and North America several times, then retreated for long periods before making new pushes to the south. A new understanding of Earth's past was born: It had been subject to a repeated series of ice ages.

The evidence left by most past ice ages is fragmentary. The continual reshaping of the Earth's surface has put such evidence through a kind of natural cement mixer. But over the past century and a half, and particularly since the 1920s, enough has been learned to indicate a general pattern of the comings and goings of the ice. An epic ice age began in the middle of the Carboniferous period, some 325 million years ago, and extended into the Permian period, 260 million years ago. That ice age was followed by a much warmer period when the dinosaurs flourished. Over the past 35 million years, ice ages have been common, occurring about every 100,000 years on average, but there have also been a number of lesser waxings and wanings. This fractured time scale means that scientists have a great deal to explain, and, as always, that opens the way to a lot of different theories and the arguments that go with them.

As scientists in different fields began to seriously consider how ice ages had occurred, there was one obvious starting point. Average global temperatures must have varied greatly over the Earth's history, and the most basic reason for such variations

This nineteenth-century drawing shows the primitive hut that Jean Louis Agassiz used with his colleagues while studying glaciers in the Swiss Alps in the 1830s. Agassiz was the first to recognize that the detritus called "drift," found all over Europe, was evidence that vast glaciers had once covered much of the continent, as well as the British Isles, during an earlier ice age. From *Louis Agassiz: His Life and Correspondence*, Vol. 1, by Elizabeth Cary Agassiz.

would have been the amount of solar energy that was reaching the surface of the planet. Even in the nineteenth century it was known that the Earth's path around the Sun is a lot less steady than it seems to us when walking down the street. It was not until the 1920s, however, that the Yugoslavian mathematician Milutin Milankovitch precisely laid out the three kinds of variation that affect the Earth's journey through space. First, the Earth travels in an elliptical orbit, not a circular one—more like the oval shape of an egg than the round shape of a baseball. In addition, there is an eccentricity even to this elliptical orbit—in the course of 100,000-year cycles, the orbit becomes less elliptical and more circular, and then moves back to the elliptical. Second, the Earth itself is tilted, and the angle of the tilt changes in the course of a 41,000-year cycle, from a maximum of 24.5 degrees away from the vertical to a minimum of 21.5 degrees. (The current tilt is almost exactly in the middle of these two extremes.) Third, the Earth also spins around its axis like a top with a wobble in it. The wobble is called "precession," and it keeps to a 22,000-year schedule. An additional small skip in the spin makes an appearance every 19,000 years.

Milankovitch spent nearly 30 years working on a series of equations that related these three kinds of eccentricities to the appearance of ice ages. He determined that at the extreme end of both the precession cycle and the tilt cycle, the amount of solar energy reaching the surface of the Earth would diminish sufficiently to allow the ice to begin expanding again. This theory made sense to many scientists, although some doubt was raised by the 100,000-year cycle affecting the elliptical orbit around the Sun. The degree of orbital change proved to be less than 0.3%, which is very small on a cosmic scale. However, it is known that the Earth's atmosphere can be affected by extremely minor factors, which is why, even with advanced computer technology, long-term weather forecasting remains a problem in respect to areas smaller than 300 miles (480 kilometers) across. Thus there was a willingness among some scientists to accept the fact that even a 0.3% change could have amplified effects on global climatic conditions.

Milankovitch's equations remained a theory, however. Some evidentiary support for his theory finally appeared in 1976, when researchers found that sediment from the sea bottom could provide a crucial indicator of the temperature of the water in past millennia. The sediment contained the shells of tiny animals called forams, and the chemical composition of the shells varied according to the temperature of the water at various eras of Earth's history. The ratio in the shells between a common oxygen isotope (oxygen-16) and a heavier and less common one (oxygen-18) varied according to the temperature of the water. The oceans, and thus the shells of the forams, contain less of the lighter isotope when the Earth's climate is cold because so much of that isotope is trapped in the glaciers forming on the surface during colder periods. Both the recovery of the sediment itself and the laboratory testing that follow are extremely painstaking and onerous work, but this work has proved enormously fruitful. The deepest layers of sediment, brought up by deep-sea drilling, have shown that the ocean depths during the Cretaceous, when the dinosaurs lived, were nearly 20 degrees warmer than they are now. That is an enormous change. Less drastic but very telling changes have been discovered, which coincide with the gradual cooling that began 115,000 years ago (when England was virtually tropical) and proceeded until the peak of the last ice age some 15,000 years ago, when the ice over southern New York was up to a mile thick—its ultimate withdrawal creating Long Island by literally dragging land into the sea.

Ice core samples, taken by western scientists drilling deep into the polar ice caps in Greenland, and taken over a long period of time by Russian scientists in Antarctica, have corroborated and extended the findings from the sea bottom. Again, the ratio of oxygen isotopes is used as a measure, and because the ice forms distinctive layers analogous to the rings one finds in trees, even more detailed data have been developed from Greenland and Antarctica to help date the warming and cooling of the Earth over the past 2.5 million years. Even though such evidence backs up the Milankovitch cycles, however, over the past few decades, a growing feeling among many scientists has been that his theory can provide only about 80%, at best, of the reasons why ice ages occur. The picture has still seemed incomplete.

The Greenland ice core samples themselves provided the clue to another major factor. In 1979, a Swiss physicist named Hans Oeschger went to Greenland to join the team of Chester Langway of the State University of New York. By crushing ice samples and collecting the gases from air bubbles trapped in the ice thousands of years ago, Oeschger was able to show that carbon dioxide levels were 100 parts per million higher when the world began warming up again about 12,000 years ago, as compared with the levels 17,000 years ago at the height of the most recent ice age. When these results were made public, new tests were carried out on the deep-sea sediments, which came up with the same results. Carbon dioxide now appeared to be the enabler that heightened the effect of solar energy cycles on the Earth's atmosphere.

How did that mechanism work? A number of important scientists have approached this problem from different directions. We know that the "greenhouse effect," so much in the news in recent years because of debates over the rate of global warming, raises temperatures. In fact, the greenhouse effect makes life on

Earth possible, with current arguments focusing on whether a rise in global temperatures will cause the global climate to spiral out of control. We also know that one of the most important factors in the process is an increase in carbon dioxide.

Why would carbon dioxide decease or increase without human intervention to upset the balance? A number of theories have been proposed for different periods in the history of the Earth. For example, the great warming that took place during the Cretaceous could well have been the result of the rapid spread of land vegetation over the face of the Earth. Such vegetation would use the carbon dioxide in the atmosphere but release it again, with the overall level increasing as more new species of plants thrived. At other times, a tremendous spurt in the amount of vegetation in the oceans could have sucked carbon dioxide down out of the atmosphere, trapping it under water and causing the kind of cooling that could tip the balance toward a new ice age.

It has also been speculated that the movements of tectonic plates, and the changes in land mass they create, could have affected the climate. In today's world, the Gulf Stream carries warm equatorial waters up the Atlantic to England, creating the relative warmth of that "green and pleasant land" at high latitudes. Perhaps the cessation of the flow of water between the Pacific and the Atlantic 2.5 million years ago with the appearance of the Central American land mass triggered glacial development in the northern hemisphere. A more recent global cooling could have been caused when Antarctica and South America separated 15 million years ago.

Another quite controversial theory is based on an erosion process in rocks, discovered by American chemist Harold Urey (for which he won the 1934 Nobel Prize). In the Urey reaction, silicate rocks draw carbon dioxide out of the atmosphere as they erode. If they are buried and regurgitated eons later in a volcanic eruption, they can release the carbon dioxide into the air again. American climatologists Maureen Raymo and William Ruddiman suggested that ice ages may have been connected to such vast mountain ranges as the Himalayas and the Andes pushing up out of the Earth and subsequently grabbing carbon dioxide out of the atmosphere as they eroded. This argument has gotten tangled up in the global warming debate, however, because these scientists go on to say that human burning of fossil fuels is acting like a volcano and releasing great quantities of carbon dioxide into the atmosphere.

In the past few years, a new theory has made headlines—at least in the science magazines—almost every year. In 1997, Richard A. Muller of the University of California, Berkeley, and Gordon J. MacDonald of the International Institute for Applied Systems Analysis in Laxenburg, Austria, made fresh use of the Milankovitch cycles in a computer model. It had already been established that 30,000 tons of cosmic "dust" fall on the Earth every year, unnoticed by us amid the general plethora of "flying smuts," to use a vivid phrase from a 1960s Nichols and May routine. Muller and MacDonald, however, theorized that the Earth passes through a particular band of cosmic dust every 100,000 years, because of the tilt of its axis, with the result that the amount of material falling to the surface of the planet increases to the point of crisis. Two other researchers, Stephen J. Kortenkamp of Washington's Carnegie Institution and Stanley F. Dermott of the University of Florida tested this hypothesis with another computer model, and they announced in 1998 that it wasn't the tilt of the Earth that mattered, but rather the shape of the orbit around the Sun—a finding more in keeping with Milankovitch's original equation. According to a May 1999 report in *Science News,* Kenneth A. Farley of the California Institute of Technology found that sediment deposits did show a threefold increase in cosmic dust every 100,000 years—but at a point when the model predicted the amount should be declining. "Something is really peculiar here," Farley concluded.

The year 1999 brought forth a new space-oriented theory, this one having to do with a dramatic increase in cosmic rays. These rays constantly bombard the Earth, but they could, in higher concentrations, cause a significant rise in cloud-cover density. Cosmic-ray bombardment can be measured using carbon-14 radioactive-decay techniques. As reported in the April 1999 *Discover,* the author of this new theory, Henrik Svensmark of the Danish Space Research Institute, was able to produce evidence that cosmic-ray activity increased by "almost a factor of two" during the last ice age.

The number of new theories being put forward concerning the cause of ice ages is, of course, a sure sign of an unsettled field of inquiry. Some scientists can get testy about some of the more exotic theories, complaining about less than rigorous computer models, and observing that self-appointed experts from far too many disciplines are butting in with sheer speculation. The very nature of the subject inevitably brings people from many different fields into the debate. In all quarters, the Milankovitch cycles are widely accepted as something that must be considered, although scientists differ about the degree of emphasis they believe should be placed on these cycles. In any case, astronomers have a secure place at the table in this debate. So do evolutionary biologists, because the forms of life that exist at a given time reflect—and may sometimes affect—what is going on in terms of climate. Geologists and chemists tend to work most closely together, as in the process of obtaining and analyzing sediment samples from the ocean deeps. While the work of teams from one discipline can sometimes bolster or even confirm that of another, there are inevitably times when they trip one another up. What might seem to be very sound reasoning from the geological point of view could run afoul of evolutionary data, and vice-versa.

In fact, numerous scientists are quite pessimistic about the possibility of ever completely solving the riddle of ice-age development. There is almost too much information pouring in, some maintain. Of course, if the Milankovitch cycles are as important as many scientists think, some sort of answer ought to appear in another 2,000 years or so. We are due for the start of a new ice age. Yet there is a problem even with that scenario. Because of the release of excessive carbon dioxide and other greenhouse gases due to human activities, the resulting global warming may have put the entire process out of whack. If that is so, we could be headed for another period of melting polar ice and nice weather for hippos in Hertfordshire (if it isn't underwater due to rising sea levels) instead of a scheduled ice age. Even that wouldn't be anything entirely new, either. After all, there was a period of more than 200 million years—dinosaur time—when there does not appear to have been anything like an ice age. Indeed, even in the past 35 million years, when there have been many glacial periods, they have not always shown up quite on time. It may be that the causes of ice ages are so varied, and so complex, that there really is no schedule, at least when the full range of Earth's history is considered. Perhaps trying to figure out this unsolved mystery is more a testament to the human need for order than anything else.

⚛ To investigate further

Lamb, Simon, and David Sington. *Earth Story*. Princeton, NJ: Princeton University Press, 1998. Based on a BBC television series, this book covers, as its subtitle announces, "The Shaping of Our World," and thus it deals with many other subjects aside from ice ages. Even so, its long, profusely illustrated chapter on the subject is thorough and well-written.

Levenson, Thomas. *Ice Time*. New York: Harper & Row, 1989. Subtitled "Climate, Science and Life on Earth," this book clearly presents the most important facts about ice-age studies, and it is not as behind-the-times as its publication date might suggest, given that the profusion of new theories in recent years are mostly at odds with one another. It also happens to be a charming and very accessible book.

Langway, C. C., H. (Hans) Oeschger, and W. Dansgaard, Eds. *Greenland Ice Core*. Washington, DC: American Geophysical Union, 1985. Readers interested in a detailed account of the ice-core drilling in Greenland, and the reasons for it, will find this book a fascinating though technical source.

Note: Readers who want to follow new developments in this field should stick to science magazines. This is one of those subjects that mass-media publications tend to sensationalize or present out of context.

Chapter 6

Were Dinosaurs Warm-blooded?

Public fascination with dinosaurs had its first full flowering in 1854, when London's Crystal Palace, an enormous iron-framed glass building originally constructed for the Great Exhibition of 1851, was reopened by Queen Victoria and Prince Albert after being moved to a new location in the suburb of Sydenham. Prince Albert suggested that the park surrounding the rebuilt Crystal Palace be enlivened with re-creations of the beasts of bygone ages, including dinosaurs. The word *dinosaur* (meaning "terrible lizard") had been coined by Richard Owen, a professor of anatomy at the Royal College of Surgeons, in a two-and-a-half-hour lecture called "Report on British Fossil Reptiles," delivered in 1841. The work of re-creating the extinct animals fell to a wildlife painter and sculptor named Benjamin Waterhouse Hawkins.

Only three species of dinosaur were known at the time: the megalosaurus of the Jurassic period, and the iguanodon and hylæosaurus of the later Cretaceous period. With no complete skeletons to work from, Hawkins had to be creative, and the results are of course mostly wrong. The iguanodon (literally "iguana's tooth") had been so named by Owen because all he originally had to go on was teeth, and they looked like those of a modern iguana. Hawkins dutifully enlarged an iguana to resemble an oversize

Richard Owen coined the word dinosaur. *In the 1860s, he posed with this dinosaur bone in front of a massive skeleton he was assembling for an unknown photographer. The background of the full photograph is badly decayed, but this detail conveys the imposing, even forbidding, character of the man.*

rhinoceros, and followed a similar approach with the other two beasts. All three were shown walking on all fours. We now know that this stance was used only by the hylæosaurus. Some lesser-known reptiles that were clearly not dinosaurs were turned into giant turtles and frogs. The sculptures were a sensation, and the iguanodon was large enough to have its back removed so that a table for 21 could be set up inside for a celebratory dinner hosted by Richard Owen. Hawkins would go on replicating dinosaurs in both England and America for the rest of his life, and because he was in fact very talented, he was able to take into account new discoveries and make some of his later creations far more accurate.

The notion that dinosaurs were all four-footed creatures was quickly dispensed with when American paleontologist and lizard expert Edwin Drinker Cope discovered the bones of what he would name *Laelaps aquilunguis* in the marl pits of New Jersey in 1866. (Laelaps was the name of a dog in Greek mythology, which the goddess Diana gave to the young hunter Cephalus; the dog was subsequently turned to stone in the act of leaping, a casualty of one of those famous Olympian squabbles among the gods.) The skeleton Cope unearthed was sufficiently complete to show that this dinosaur had walked on two legs, and that its front legs were so short that they resembled small arms. This find established that at least some dinosaurs had leaped about like gigantic kangaroos.

The most lasting of the assumptions based on the dinosaur/lizard connection, however, was that they had to be cold-blooded animals like crocodiles, rather than warm-blooded ones like mammals. Cold-blooded animals are termed *ectothermic,* meaning that they must absorb body heat from the sun. Warm-blooded animals, including all mammals, are *endothermic,* generating their own internal heat. This does not mean that lizards are literally cold-blooded, however; their body temperatures can be as warm as those of mammals and sometimes even warmer. They simply have entirely different biological systems for regulating body temperature.

All amphibians and most reptiles have three-chambered hearts, with two thin-walled atria that expand to receive blood, and one thick-walled ventricle that pumps the blood back out. Birds and mammals have four-chambered hearts, with two atria and two ventricles. The single ventricle of a lizard's heart has to perform a double function, not only pumping the blood from the lungs into the body, but also recycling the body's blood back into the lungs. The lungs replenish the oxygen supply in the blood, and that oxygen is depleted as it travels through the body and back to the lungs again. In the lizard's single ventricle, both the freshly oxygenated blood from the lungs and the "used" blood that has circulated through the body inevitably get mixed together, meaning that there is much less oxygen-rich blood to be pumped to the muscles where it provides the energy for activity. In order to

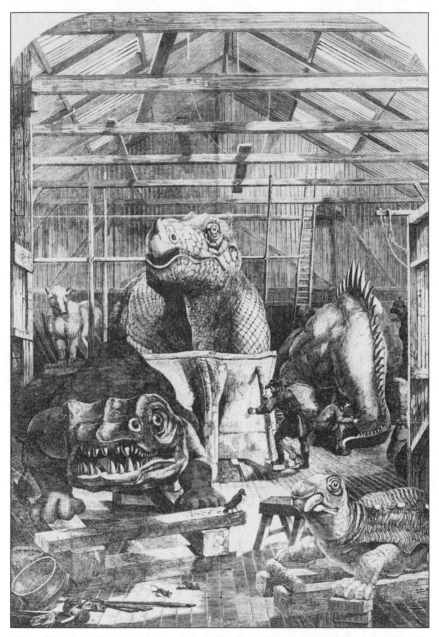

This etching from 1854 depicts the workshop of Benjamin Waterhouse Hawkins, who designed and built the fanciful dinosaur figures commissioned for the grounds of the Crystal Palace when it was moved to the suburbs of London in 1854. Queen Victoria visited the studio with her husband Prince Albert; the creation of the Crystal Palace had been his idea. The creature at center rear had a removable back, and Richard Owen, Hawkins, and a dozen dignitaries held a celebratory dinner at a table placed inside the sculpture. From *Dinosaurs*, by W. D. Matthew.

acquire that energy, lizards and crocodiles must sun themselves for long periods (as much as 90% of daylight hours) to absorb enough heat to make sustained activity possible. In mammals and birds, the second ventricle of the heart keeps separate the oxygenated blood from the lungs and the used blood from the rest of the body, and the muscles receive a far greater supply of oxygen. The only drawback to the mammalian and avian system is that it requires greater quantities of food to keep it running.

Despite the realization 130 years ago that some dinosaurs walked on two legs, and increasing skeletal evidence that these creatures were constructed to move very fast, despite the fact that the jaws and teeth of many dinosaurs were adapted to a voracious consumption of meat that would give them the energy for such rapid movement, scientists clung to the idea that dinosaurs had to have been cold-blooded animals like modern reptiles. This belief persisted despite a growing conviction that birds were descended from dinosaurs, making it necessary to explain why birds should have four-chambered hearts that allowed for a great degree of activity in their everyday lives and also made it possible for them to undertake migrations across thousands of miles. Of course, birds did lay eggs, like dinosaurs and unlike mammals, which provided an excuse for ignoring the discrepancy between modern bird and lizard hearts. From time to time, some brave souls would suggest there might be a real problem here, but they were always quickly cowed into silence.

Until 1969—finally, someone with too much authority to be ignored began arguing that at least some dinosaurs, certainly those that walked on two legs, must have been warm-blooded creatures with four-chambered hearts. John Ostrom, a professor at Yale University, had discovered the dinosaur Deinonychus, which used its clawed hind feet to kill prey, requiring extraordinary balance and agility. In 1969 he delivered a lecture at the first North American Paleontological Convention on the subject of what could be inferred from fossils regarding the climate in the Mesozoic period.

This innocuous-sounding discourse contained a major intellectual bombshell. Because dinosaurs had roamed in the high latitudes of the Earth during the Mesozoic, and because dinosaurs were regarded as cold-blooded, it had long been assumed that the climate must have been tropical as far north as present-day Canada. Otherwise, the cold-blooded dinosaurs wouldn't have been able to absorb sufficient heat to keep going. While ostensibly questioning such ideas about climate, Ostrom made it clear he did not buy the entire notion that the gigantic dinosaurs spent 90% of their time lying around in the sun in what is now Wyoming. A great many of them walked erect after all—which lizards do not and cannot do because even if they had the legs to support themselves, they wouldn't have enough energy. "The correlation of high body temperature . . . high metabolism and erect posture and locomotion is not accidental," Ostrom said. Then came the kicker: "The evidence indicates that erect posture and locomotion probably are not possible without high metabolism and high uniform temperature." In other words, the dinosaurs were warm-blooded.

A few scientists were quick converts to the idea that many dinosaurs had to have been warm-blooded. One of those was the paleontologist and writer Adrian J. Desmond, who published *The Hot-blooded Dinosaurs* in 1976. After discussing Ostrom's speech, he wrote, "Nobody before had demonstrated the inextricable relationship between high metabolism, stable temperature, and erect posture, yet once explicitly stated this linking seemed obvious and natural. It resolved the long-standing contradictions inherent in the ludicrous sun-basking brontosaurus model by scrapping the model altogether and substituting an endothermic dinosaur. Of course, this requires a radical reappraisal of dinosaurian physiology; and we are compelled to look at the mammal and the bird for our new model."

A former student of Ostrom's, Bob Bakker, expanded the argument for endothermic dinosaurs 11 years later in his 1986 book, *Dinosaur Heresies,* insisting that the very success of the dinosaurs depended on their being-warm blooded. But note his title. While Desmond had dared to put this claim in his title in 1975, Bakker felt compelled to note the strength of the opposition in his. Nevertheless, Bakker boldly argued that not only did the upright posture of many dinosaurs suggest that they were endothermic, but also because they often carried their heads high in the air on long necks, it would take the high blood pressure associated with warm-blooded animals to move blood to their brains. Scientists on the cold-blooded side weren't impressed with that argument. Giraffes, they noted, have special valves in their neck, which help move the blood upward; maybe dinosaurs did, too. Yes, came the reply from the warm-blooded camp, but giraffes are mammals, remember?

This kind of back-and-forth sniping continued right through the 1990s. The arguments against dinosaurs being warm-blooded were laid out with a good deal of clarity in a 1995 book published under the banner of New York's renowned American Museum of Natural History: *Discovering Dinosaurs,* by Mark Norell, cocurator of the museum's Hall of Dinosaurs; Eugene Gaffney, cocurator with Norell, as well as curator of vertebrate paleontology; and Lowell Dingus, director of the museum's Fossil Halls renovation. They adopted a rather cautious middle path in this debate, but they generally marshaled more evidence against the *endothermic* (warm-blooded) theory than for it. For example, they discuss in some detail the analysis of bone microstructure, which differs between *ectothermic* (cold-blooded) and endothermic animals that are now living. They report that the bones of dinosaurs have been tested by cutting them into very thin slices and then comparing them under the microscope with the bones of modern animals. "In most nonavian dinosaurs, the microstructure of the bone appears more like the bone of endothermic animals"—the warm-blooded argument—"but this evidence is not conclusive."

They go on to say, "To standardize the observations, we must compare animals of similar size. Unfortunately, no nonavian, dinosaur-sized ectothermic animals are alive today, and we have only begun to study small dinosaurs." This is a curious statement. After all, the original idea that dinosaurs were cold-blooded was based on the fact that vastly smaller living reptiles such as iguanas and crocodiles are cold-blooded. Why is it all right to extrapolate from smaller creatures when supporting the cold-blooded theory, but not all right when supporting the warm-blooded theory? Nonetheless, the authors manage a last sentence in this section that is a masterpiece of equivocation and at least leaves open the possibility that dinosaurs were endothermic: "There is no clear-cut evidence that dinosaurs were either cold-blooded or warm-blooded, except that dinosaurs evolved endothermy sometime in their history, as documented by living birds."

As this last sentence indicates, there was now at least enough doubt so that these experts regarded both theories unproved. That was certainly progress, but perhaps not enough when one considers that the word *dinosaur* was coined in 1841. In that regard, it is worth going back and taking a look at why Richard Owen was so determined that dinosaurs should be cold-blooded in the first place. Certainly, it made some sense from the start— after all, Owen had named the iguanodon on the basis that its teeth were very much like those of the modern iguana. In addition, as Adrian Desmond discusses at some length in his 1975 book, Owen also had an ulterior motive. He was a highly religious man, and he had been disturbed by the early rumblings of evolutionary theory that had come from French naturalist Jean Baptiste de Lamarck at the very beginning of the nineteenth century. Lamarck, who coined the word *biology* and was the first to distinguish between vertebrate and invertebrate animals, believed that organisms had an intrinsic drive or urge to evolve into better-adapted ones. Although initially influenced by Lamarck, Darwin would make clear that evolution was accidental rather than purposeful. Owen didn't like either of these theories because they both seemed to downgrade the role of the Creator. When Owen proclaimed that dinosaurs were prehistoric lizards, however, Darwin's *Origin of the Species* was still 15 years down the road. It was Lamarck whom Owen initially wanted to undercut. If gigantic lizards had once roamed the Earth, and now we had only small ones, it seemed to show that Lamarck's belief in self-improving evolution was hogwash. The great lizards were long gone, and only much smaller creatures descended from them were left. So much for the notion of self-improvement!

The ideological basis for Owen's insistence on the cold-bloodedness of dinosaurs did not prevent the concept from ruling the roost for more than a hundred years. Even when a real scientific attack on this notion was finally launched, a great many distinguished scientists were loath to let it go. Equivocation was as far as the prestigious scientists from the American Museum of Natural History were willing to venture in 1995. Other experts even have a partial answer to the ridiculous image of a *Tyrannosaurus rex* (or the vicious velociraptors Steven Spielberg had so much fun with in *Jurassic Park*) lying around in the sun for eight hours absorbing sufficient energy to go look for a meal. Animals of such enormous size, they say, would not lose heat the way a crocodile does, and thus they wouldn't have to spend so much time sunning themselves. Still others take the opposite tack, suggesting that if dinosaurs had really been warm-blooded they would have overheated and needed to take baths to cool off the way elephants do. This intellectual merry-go-round can get a little dizzying, but there always seems to be a way to keep it going. No evidence ever seems quite good enough for everybody. It is, of course, the duty of any scientist to insist on hard evidence. But as we have seen in earlier chapters, and will see again, that rule doesn't always apply if the stakes are high enough.

Ever since John Ostrom dared to disrupt the old consensus in 1969, scientists on both sides of this debate, as well as those trying to keep a solemn face while occupying the middle ground, had been saying that the questions could be answered if only a dinosaur heart could be found. That discovery seemed impossible— nothing but bones had ever been discovered. Then, to almost everyone's astonishment, it was announced in mid-April of 2000 that a fossil dinosaur heart had indeed been found, in the chest cavity of a dinosaur skeleton uncovered in South Dakota. Encased in stone, the heart was about the size of a grapefruit.

The discovery was announced by Dale A. Russell of the North Carolina Museum of Natural Sciences in Raleigh. The internal structure of the heart, of which there were visible traces, indicated that the organ was more like that of a bird or a mammal than anything seen in a reptile before. "The implications completely floored me," Russell said. He also said that while it seemed to be a four-chambered heart, a great many tests would have to be done over the next few years to make certain. As a start, the stone containing the heart had been scanned using special computer software that took two-dimensional images and turned them into a three-dimensional model. Two ventricles and the aorta could be seen, according to the investigating team, but the upper chambers, or atria, were not visible.

What made the discovery even more important was the fact that the dinosaur in which the heart was found was not of the lineage believed by most paleontologists to have evolved into birds. The dinosaur itself was a plant-eating animal judged to have weighed about 600 pounds (1323 kilograms) and to have a length of 13 feet (4.5 meters), one of the smaller dinosaurs that lived about 65 million years ago just before the mass extinction discussed in chapter 3. Its exact species has not been determined, but it was a member of the genus known as Thescelosaurus, or "marvelous lizard."

With this new evidence, Mark Norell, one of the authors of *Discovering Dinosaurs,* seemed to come down off the fence a bit, telling John Noble Wilford of the *New York Times,* "This means our entire conception of dinosaurs may have to be revised." Doubt still lingered in other quarters, however. University of Chicago paleontologist Paul C. Sereno told Wilford that he had "serious reservations," questioning whether internal organs could have been preserved in the kind of sediments usual in the area where it was found on a rancher's land. He also said that the images he had seen didn't provide clear enough evidence that the organ was even a heart.

In other words, an organ that some experts say is a dinosaur heart, and a four-chambered one to boot, is only a "putative" heart to another expert. A putative heart is not going to prove that the dinosaurs were warm-blooded. Also, it's a good bet, given the nature of this particular debate, that if the putative heart is finally shown to be an actual heart, someone will say that this is just one dinosaur from a very late period when birds were already evolving, and that it doesn't prove anything about other dinosaurs.

Sometimes one gets the impression that there are certain mysteries some scientists prefer to have remain unsolved.

❊ To investigate further

Desmond, Adrian J. *The Hot-blooded Dinosaurs.* New York: Dial Press, 1976. Despite having been written more than 25 years ago, this is a fascinating and rewarding book. Desmond holds a degree in the history and philosophy of science, as well as in vertebrate paleontology, and this book is as interesting for its historical account of dinosaur studies as it is for its title thesis. It does not have the kind of lavish color drawings of dinosaurs that some readers look for, but its wonderful old prints will delight others.

Bakker, Robert. *The Dinosaur Heresies*. New York: Morrow, 1986. Bakker lays out the most detailed case for the warm-blooded theory, and his work may yet prove to be ahead of the game.

Norell, Mark A., Eugene S. Gaffney, and Lowell Dingus. *Discovering Dinosaurs*. New York: Knopf, 1995. Despite its caution on the cold-blooded versus warm-blooded controversy, this is an excellent book focusing on the fossil record, well illustrated with photographs of actual fossils, as opposed to Spielberg-esque speculations in glorious color. It is organized around 50 specific questions about dinosaurs and is therefore excellent for dipping into or for finding the answer to a specific question such as "How did dinosaurs mate?" or "How large were the biggest dinosaurs?"

Lambert, David. *The Ultimate Dinosaur Book*. New York: DK Publishing, 1993. For those looking for visual splash, this is a typically lavish Dorling Kindersley production, with plenty of facts, as well.

Stevenson, Jay, and George R. McGhee. *The Complete Idiot's Guide to Dinosaurs*. New York: Alpha Books, 1998. Like many other books in this humorously titled series, this oversize paperback is designed to make you feel much smarter after reading it. Although broken up into small bites of information for easy absorption, it contains a great deal of information and several extremely useful appendices.

Chapter 7

Is There a Missing Link?

n 1856, three years before Charles Darwin published *On the Origin of Species,* parts of a skeleton were found near Düsseldorf, Germany, in an area called the Neander Valley. They caused a great deal of speculation because they seemed quite peculiar. The only person at the time to recognize the find as being something other than human bones, however, was the English anthropologist William King. He coined the name *Homo neanderthalis* for what he believed to be a different kind of *hominid* (any two-legged primate, including the apes and the precursors to modern man). The name stuck, but even he changed his mind about the bones belonging to a separate species, and it would be half a century before that concept became widely accepted.

It would later become clear that other such bones had been found earlier, but that their significance had been ignored. King's original belief that they represented a different kind of hominid from human beings received a great deal of attention because Darwin's basic ideas about evolution, as well as those of his rival Alfred Russel Wallace, were already sufficiently known in scientific circles to have stirred the beginnings of a debate that continues to this day. Then, as now, there were those who reacted with horror to the idea that human beings were even related to the apes, finding that concept an affront to both God and humanity. The Neanderthals (or Neandertals, as Donald Johanson, who discovered the famous "Lucy" skeleton in 1974, and some other anthropologists prefer) came to be regarded by many scientists as "brutish." In the nineteenth century, even professional scientists seemed to be infected to some degree by the distaste that the religious had for the idea that we *Homo sapiens* might be intimately linked to "bestial" creatures.

As Johanson and others have pointed out, the degrading view of the Neanderthals that persisted into the 1950s can be blamed on one man, French anthropologist Marcellin Boule. Boule declared that these primitive brutes could in no way be compared to the Cro-Magnons, who settled in Europe 35,000 years ago and are generally regarded as the earliest human beings. The first remains to be called Cro-Magnon were discovered in the Dordogne region of France in 1868. Boule saw the Neanderthals as subhuman, but he described Cro-Magnons as having "a more elegant body, a finer head, an upright and spacious brow, and who have left, in the caves which they inhabited, so much evidence of their manual skill, artistic and religious preoccupations, of their abstract faculties, and were the first to merit the glorious title of *Homo sapiens!*" These words were written in 1908, following the discovery of a deformed Neanderthal skeleton that we now know to have been twisted by arthritis—to which the Neanderthals were prone.

The scientific community in general went along with Boule's belief that we could not have evolved from this brute stock. Nonetheless, it was clear that, given the slow workings of evolutionary change, there must be some creature, further back, that could stand as an intermediary between the apes and our exalted selves. Thus was born the concept of the "missing link," and thousands of amateur geologists went out hunting for bones that might turn out to be significant. In the late 1800s and early 1900s, such enthusiasts played the kind of role that amateur astronomers have in the search for new comets in recent years. In 1912, one such man, a British lawyer named Charles Dawson, found what appeared to provide the answer. In a gravel bed on Piltdown Common near Lewes, England, he uncovered a skull with a cranium that was clearly human, but which also had an apelike jaw.

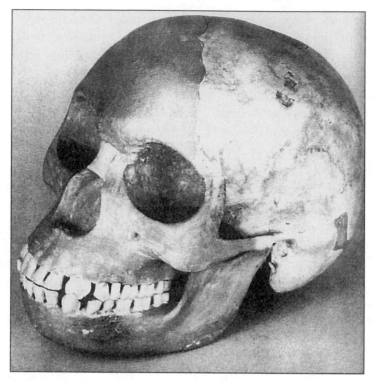

A photograph of the supposed skull of Piltdown Man, discovered by Charles Dawson in England in 1912 and certified by numerous experts as authentic evidence of a "missing link" in human evolution. It was exposed as a fraud in 1953. Courtesy the Museum of the City of London, England.

Piltdown Man, as it came to be called, created a worldwide sensation. The skull was put through a variety of tests by top scientists, and they declared it authentic. There was some concern that no other bones could be found in the area, but theorists have always had a knack for cooking up excuses for anomalous finds, and plenty of them were put to use in this case. Piltdown Man went into the biology books as the proven missing link, the certain answer to those who claimed that humans could not possibly be related to the apes. There were always a few anthropologists who were dubious about Piltdown Man, and eventually their worries prompted a new round of tests in 1953, using chemical analyses that had been recently developed. The headlines that had once screamed "MISSING LINK FOUND" now read "PILT-

DOWN HOAX." It had been shown that the skull consisted of a human cranium attached to an orangutan jaw. The joining had been performed with enormous skill—but perhaps not quite enough to excuse 41 years of abject credulity.

It then took another 43 years to determine who had perpetrated the hoax, one of the most successful and damaging in the history of science. Over the next several decades, the puzzle attracted numerous researchers, who came up with all kinds of suspects. Poor Charles Dawson, having discovered the skull, was inevitably a popular culprit, but no one was able to demonstrate that he possessed the skills required for creating the human/ orangutan assemblage. Finally, in 1996, two British paleontologists solved the mystery after nearly a decade of work. Brian Gardiner and Andrew Currant uncovered crucial evidence in an old trunk moldering away at the British Museum. As reported in *Discover* magazine, the trunk contained bones that had been dipped in acid and treated with manganese and iron oxides to age them— just as the Piltdown skull had been. The trunk was inscribed with initials M.A.C.H. These matched up with a man who had been the keeper of zoology at the British Museum in the 1930s and 1940s, whose last name was Hinton.

What possible motive could Hinton have had? Further investigation revealed that he had started working at the museum as a volunteer in the first decade of the twentieth century. He had had the temerity to ask for a salary and been sneeringly refused by the then keeper of paleontology, Arthur Smith Woodward. As Woodward was bound to be called on to examine a find such as the Piltdown skull, its creation could serve as a trap to embarrass him. Indeed, Woodward was one of those who authenticated it. By this time, however, Hinton was well on his way to becoming a respected scientist himself. To reveal the hoax would have caused him as much trouble as it would Woodward. It has been speculated that Hinton left the trunk at the British Museum in the hope that it would eventually be recognized for what it was, and even during his lifetime he planted another clue. Among the interests listed in his entry in the British *Who's Who*, he included "hoaxes."

By 1953, when Piltdown Man was exposed as a forgery rather than the missing link, the study of the hominid evolutionary chain stood ready for revision anyway. The Boule concept of Neanderthals as hairy brutes had been falling apart for some time, and many scientists were ready to acknowledge that they might be very closely related to us. Thus the experts quickly embraced a new view put forward at a 1956 symposium by the anthropologists William Straus and A. J. E. Cave. Their analysis of the bones discovered in 1908, on which Boule had based his derogatory conclusions about Neanderthals, showed the presence of arthritic disease, and from other finds, it was evident that healthy Neanderthals had stood fully erect instead of slouching about like apes. Their report, published in the *Quarterly Review of Biology,* went so far as to suggest that a cleaned-up, properly attired Neanderthal could go unnoticed on a New York subway train—although some wits promptly took this as an insult to New Yorkers.

Over the next three decades, the majority of scientists came to the conclusion that instead of being some primitive hulk, the Neanderthals were quite probably our own most immediate ancestor. New fossil finds made it clear that the Neanderthals had made tools, used fire, and indeed appeared to have had bigger brains than we do. Still, there were doubters. While it was generally accepted that Neanderthals were far more advanced than had been believed in the first hundred years following the Neander Valley discovery, some problems remained. One of the main ones had to do with the lack of anatomical evidence that the Neanderthals could speak. The larynx seemed to lie too high to be conducive to anything more than the noises uttered by chimpanzees, although it is recognized that even such constricted grunts and screeches can convey a considerable amount of information to other chimpanzees—more, according to some field researchers, than we like to admit. In the summer of 1983, however, the discovery of a nearly intact male Neanderthal skeleton in the Qafzeh cave in Israel provided evidence that most scientists had despaired of ever finding. It was a delicate U-shaped hyoid bone, which in human beings is attached to the cartilage of the larynx. That indicated speech—and a talking Neanderthal would be an even better candidate for our direct ancestor.

Two years earlier, in the magazine *Science '81,* anatomist and illustrator Jay Matternes had published a portrait of a Neanderthal male (developed by using plaster casts of fossils) that was astonishingly human in appearance, despite its bulbous nose and heavy browridge. There were even those who suggested that the bald-headed figure looked a lot like Pablo Picasso. This famous article (available on the Internet at www.bearfabrique.org/Evolution/neander), the subsequent discovery of the hyoid bone, and the fact that Neanderthals had been found as far afield from western Europe as Israel, all contributed to the growing conviction that these hominids were our immediate predecessors. Even Donald Johanson "shared the sentiment" that Neanderthals belonged to our own species when he wrote *Lucy: The Beginnings of Humankind* in 1981. But he has since changed his mind, for reasons that illustrate both how fast things are changing in anthropology and why there is still so much disagreement.

"Lucy," the 40% complete skeleton of a young female discovered by Johanson and French colleagues in 1974, has become the most famous find since Piltdown Man—except that there is not the slightest doubt about Lucy's authenticity. Uncovered at Hadar, in the Afar region of Ethiopia, where fragments of an additional 13 skeletons were found in 1975, Lucy and her family, who may have perished in a flash flood, quickly seized the public imagination, for several reasons. They are the only fossils of prehumans that exist from the period 3–4 million years ago, and Johanson succeeded in persuading most anthropologists that they represented the stem species from which all subsequent hominids developed. This led the press to refer to Lucy as the "mother of us all," a concept that carries with it the kind of romantic mystery certain to appeal to a broad public. Does this mean, however, that Lucy and her kind, which Johanson called *Australopithecus afarensis,* are the "missing link" between the apes and humankind? The answer to that question is complex, and in the end, it can be answered either with a "maybe," or with another question: What do we really mean by a "missing link"?

The term *Australopithecus* is used to describe an entire genus of hominid creatures made up of at least five different species. Lucy's species, *afarensis,* dates to at least 3.5 million years ago, but some of the others came much later, with the entire genus becoming extinct about 900,000 years ago. The matter is complicated by the fact that anthropologists have come to divide the Australopithecus genus into two distinct groups: the *gracile* and the *robust*, words implying just what anybody might think. In his book *Ancestors,* Johanson notes that the names are misleading because they call to mind "ballet dancers versus wrestlers," when in fact they were all probably quite similar in overall body size. While that similarity can't be established with absolute certainty because the major fossil evidence consists mostly of skulls and teeth, those very skulls, and the indication of brain size and facial characteristics they provide, is the most telling kind of evidence. By that yardstick, the Australopithecus finds in Africa clearly divide into two gracile species (Lucy's *afarensis* and a later *africanus*) and three robust species.

Not all anthropologists are happy with this emphasis on skulls, however, feeling that it tends to obscure other crucial differences. A polite but intense debate developed, for example, between anatomist Owen Lovejoy, who made the first analysis of Lucy's bones, and a team from the State University of New York at Stony Brook headed by Randy Susman. As Ian Tattersall of the American Museum of Natural History in New York notes, Lovejoy thinks of Lucy as "a perfectly adapted biped" that lived on the ground and walked upright, while Susman and his colleagues "point to the rather long, slightly curved hands and feet as evidence of arboreal activity, suggesting that these humanoids habitually slept in trees for safety, and perhaps got a lot of their food there, too." The Susman point of view suggests that Lucy and her species were more primitive than Lovejoy—and Johanson—believe. This debate has been argued in almost excruciating detail since the early 1980s, to the point that the lay reader might be forgiven for saying "so what?" But the underlying implications of the debate have consequences. If Lucy was closer to the apes, then her credentials as "the mother of us all" suffer to a degree, even as the possibility that she is indeed the missing link become more persuasive. These implications are skirted by both sides, however, for the simple reason that other hominids further down the line introduce other disagreements that are even more important.

In the course of the twentieth century, our knowledge of human precursors and early humans has greatly expanded as a result of major discoveries of hominid bones, both human and prehuman, including the Peking Man fossils found in China in the 1920s, the Java Man fossils unearthed on that island in the late 1930s, the Mungo fossils found in Australia beginning in 1968, and finds in Israel from the 1960s onward, including the nearly complete Neanderthal male discovered in 1983. These finds have also led to a whole new controversy, however. The finds in Africa during the 20th century—including those of the Leakeys (Louis, his wife Mary, their son Richard, and their daughter-in-law Meave) at Olduvai Gorge in Northern Tanzania near the Serengeti Plain, the discovery of Lucy at Hadar in Ethiopia; and the earlier discoveries of Raymond Dart at Tuang in South Africa—have convinced almost everyone in the field that hominid development up to the point of *Homo erectus,* who lived in Africa about a million years ago, took place entirely on that continent. From that point on, however, opinion divides into two separate camps.

The "out of Africa" camp holds that not only did all earlier hominid species develop in Africa, but that the first of our own kind, *Homo sapiens,* first appeared there, between 500,000 and 100,000 years ago, and subsequently spread around the world, moving north into Europe, east through what is now Israel and Iraq on into Asia, and eventually by raft to the islands of the South Pacific and then Australia. The other point of view, known as "multiregional," holds that it was *Homo erectus* that made the great journey out of Africa to the rest of the world, and that *Homo sapiens* then developed independently in many parts of the globe, with different racial characteristics that were determined by varying environmental conditions.

There is no argument about the fact that *Homo erectus* did travel out of Africa—its remains have been found in many places, from China to Australia. The species was superbly equipped for making long journeys, being taller than we are and having skeletal differences that would have given its members enormous speed and stamina on foot. Even given the fact that *Homo erectus* was able to traverse the globe over a period of thousands of years, that fact does not, in and of itself, prove that the species evolved into modern humans every place it went. The multiregional camp insists that that is exactly what happened, but the out-of-Africa camp says, no, modern humans evolved from *Homo erectus* only in Africa, and then spread out in the same way that *Homo erectus* had previously done, displacing that earlier species, as well as the "archaic" humans that had evolved from it in some places.

The evidence marshaled by each side in this debate can be very convincing when each is read by itself. But when point-by-point rebuttals are made, holes begin to appear on both sides. The balance, as things stand now, may be tipped by finds in caves at the Klaises River mouth on the Cape coast of South Africa. Several different dating techniques were used to determine the age of these fossils, which suggest beings that looked very much as we do today, and the results put their age at 75,000–115,000 years ago. There are no other anatomically modern human (*Homo sapiens*) remains that go back that far anywhere else in the world. While some fair-minded proponents of the out-of-Africa thesis are willing to say that *Homo sapiens* fossils of such great age may exist in China or Java, unless or until they are found, the possibility that modern humans arose only in Africa remains the most convincing argument. It should be noted that the new field of "evolutionary DNA analysis," which is controversial in itself, clearly sides with the out-of-Africa proponents.

There is another attraction to the out-of-Africa thesis. It allows the clearest line of evolutionary descent to be drawn down from Lucy and her *afarensis* family, starting about 3.5 million years ago to the presence of modern humans approximately 100,000 years ago, all on one continent. Lucy, in this view, stands as a "great-grandparent" to modern humans, although she may well have also been the "parent" of other side branches that eventually withered. About a million years ago, our most immediate parent, *Homo erectus,* appeared. In this picture, Lucy herself stands as the "missing link" between the apes and the hominid species that led ultimately to our own development. There is another mystery still to be considered, however, one that suggests a different kind of missing link entirely.

Here, we must return to the story of the Neanderthals. As related earlier, this species of hominid was regarded as a brutish nonhuman for nearly a century, and then, in a complete reversal in the 1950s, was enshrined as our most immediate ancestor, falling between *Homo erectus* and ourselves. But that view, although still held by some, ran into serious trouble in 1988 when a new dating technique called thermoluminescence (TL) was developed by French archeologist Helene Valladas, working at the Center for Low-Level Radioactivity in Gif sur Yvette, France. Carbon dating, itself introduced only in the 1950s, had allowed archeologists to fix dates back to about 40,000 years ago by measuring the radio-active decay of fossil finds and the surrounding terrain, but TL made it possible to go back as far as 300,000 years in some cases. It also proved the contention of archeologists Ofer Bar-Yosef and Bernard Vandermeersch that evidence at various sites in Israel suggested that modern humans and Neanderthals were living in that area of the world *simultaneously,* and that they had probably interacted.

In the 1990s, it was established that this was in fact the case, and that Neanderthals appeared to have learned from humans (a.k.a. *Homo sapiens*) how to make more sophisticated stone tools than they had developed on their own. In addition, it became clear that humans had lived in the area before Neanderthals did, as much as 90,000 years ago. That punched an enormous hole in the idea that the Neanderthals were our immediate ancestor. Although early Neanderthals go back much further, existing at least 180,000 years ago, Neanderthals and humans of the Cro-Magnon period had overlapped by tens of thousands of years, and they could only be seen as separate species. To what degree they interacted is a matter of conjecture.

Ian Tattersall begins his 1995 book *The Last Neanderthal* by imagining two very different fictional scenes. In the first, he has an aged Neanderthal woman (meaning in her early 40s because we know of few Neanderthals who lived much beyond that age) watching her "human" grandson kindle a fire. He is the changed result of her own youthful agreement to become the mate of one of the tall "intruders" who had appeared out of nowhere and whose genetic stock had prevailed over her own. In the sec-ond imagined scene, "the last Neanderthal" is a male being hunted down by a group of these intruders, whom we call Cro-Magnons. These scenarios represent two schools of thought. Did the early humans interbreed with Neanderthals, and drive them to extinction through genetic superiority, or did they violently exterminate them?

A third possibility is now gaining favor. It seems possible to many archeologists, anatomists, and experts on the extinction of species that in many places, including the areas of Israel discussed previously, the two different species coexisted for thousands of years, unable to produce viable offspring but living in relative harmony so long as there was enough food to support both groups. In this scenario, the Neanderthals died out simply because they were not as smart, and because their limited life span meant that they could not produce as many offspring as the longer-lived Cro-Magnons. No matter what happened, most experts emphasize that the Neanderthals were the most advanced hominid on Earth for about twice as long as modern humans have existed, and they should in no way be disparaged. We can start crowing, in other words, in another hundred thousand years.

Despite all the fossil evidence that has been uncovered in the past century, despite our greatly increased knowledge about the evolutionary developments that led to the rise of us *Homo sapiens,* we remain in the dark about long periods of the 3.5 million years since Lucy lived. What we do know pales beside what we do not know. New archeological digs and further refinements in dating techniques and DNA research may tell us even more, but it seems likely that numerous "missing links" will remain—not merely in the sense of a real skeleton like Lucy's, or a fake skull like that of Piltdown Man, but also in terms of a true understanding of what changes took place in the long succession of hominid beings, many of them dead-end strains and some forever lost to us, leading to modern human beings.

Why did hominid creatures stand up and start walking on two legs in the first place? Experts can speculate, but we do not really know. Were the forests, or jungles, of Africa shrinking? Was the spreading Serengeti Plain extremely inviting, offering kinds of food that required a different sort of locomotion to take advantage of? Did mating roles or the care of offspring play some part? Was it just a freak development, a kind of evolutionary accident? After all, walking on two legs is, as Donald Johanson has said, "one of the oddest behaviors found in nature." Also, it presented all kind of problems, some of which we are still paying for with backaches. More profoundly, it meant that the pelvic region had to be reshaped, and become smaller, even as larger brains were needed to cope with the bipedal world.

Eventually, that meant that infants had to be born with less developed brains or their heads would not be able to pass through the birth canal. Human brains more than double in size in the first year following birth and do not reach adult size until age six or seven. That means that human children must be cared for much longer than is true of the apes. That, in itself, would lead to the even larger brains necessary for giving such care, and a more sophisticated kind of social organization to create the support structure required.

Standing erect thus forced further evolutionary changes, which led to still others. After nearly 4 million years of such changes, true humans began to appear. Through all that time, at every step, there are mysteries that run far deeper than the mere physical chain that archeologists dig in deserts, caves, and swamps to uncover. These changes are beginning to be seen as a more profound—and more elusive—kind of missing link than any skull or skeleton can provide answers for.

The greatest mystery is what happened some 100,000 years ago, when the Neanderthals, with their fires and primitive tools, had long been the most advanced being on Earth. Along came another species, whether out of Africa or from next door, with greater speech capacity and the ability to make ever better tools. Even those developments, however, were not nearly as significant as the sudden appearance among these new hominids, humans we call Cro-Magnons, of the urge to create art. There are objects, carved horses, that go back 32,000 years, and cave paintings such as those at Lascaux, France, which date back 17,000 years. The Neanderthals had fire and tools, and they appear to have been capable of making better tools after seeing what humans were up to—but they did not have art. Art was the start of a new way of thinking, of making symbols that would eventually lead to written languages, and to the beginning of recorded human history. Some new connection in the brains of humans, some linkage that the Neanderthals did not have, brought the dawn of civilization. How did that linkage occur? We do not know, and few scientists believe we will ever really know. That, ultimately, is the true missing link.

⚛ To investigate further

Johanson, Donald, Leonora Johnson, and Blake Edgar. *Ancestors: In Search of Human Origins.* New York: Villard Books, 1994. This companion volume to the Nova television series (itself available as a video through WGBH in Boston) is at once accessible and detailed, covering the field with both breadth and fairness. While Johanson (despite his collaborators, the book is written in the first person) quite naturally emphasizes his own conclusions about the development and spread of *Homo sapiens,* he gives those with other views ample chance to speak for themselves.

Johanson, Donald, with James Shreeve. *Lucy's Child: The Discovery of a Human Ancestor.* New York: Simon & Schuster, 1981. While some of the information in this book is outdated due to more recent discoveries (including Johanson's own), the unearthing of Lucy and her family was one of the most important events in the history of anthropology, and this detailed account of it makes for fascinating reading.

Tattersall, Ian. *The Last Neanderthal.* New York: Macmillan, 1995. This large-format, profusely illustrated book, subtitled "The Rise, Success, and Mysterious Extinction of Our Closest Human Relatives," is both entertaining and packed with information. Tattersall, chair and curator of the Department of Anthropology at the American Museum of Natural History in New York, certainly knows his stuff, although he does not give as full an account of disagreements in the field as Johanson does in *Ancestors.*

Jolly, Alison. *Lucy's Legacy.* Cambridge, MA: Harvard University Press, 1999. For those looking for a different take on evolution and the rise of human beings, Jolly, one of the world's foremost primatologists, offers a view that stresses cooperation rather than "survival of the fittest" as a crucial element in evolution, particularly in regard to primates. This book has been widely praised for its wit and its emphasis on the contributions of the females in primate species.

Note: Because there are so many debates about anthropological issues and human evolution, and because new technologies and avenues of research are constantly appearing, readers with a special interest in the subject should keep their eyes peeled for newspaper and magazine stories about the latest developments. It may be, for example, that DNA researchers will trump the archeologists in the not-too-distant future.

Chapter 8

What Caused the "Big Bang" in Human Culture?

At the Pech-Merle cave at Lot, France, there is a painting of a horse on a rock face, the shape of which naturally suggests the outline of the animal. It is surrounded by large blobs of paint and the stenciled outlines of hands. Precise dating is impossible, but it has been established that it was painted during the Aurignacian period, 35,000 to 25,000 years ago. The ivory head of the "Venus" of Brassempouy, Landes, France, was carved between 27,000 and 22,000 years ago. Carvings, engravings, and paintings that date from that time forward have been found all over the world, even though everywhere except Europe the engravings and paintings were created in the open air, obviously cutting down on their chances of preservation. Although these artistic remnants of hunter-gatherer societies go back a long way, it has become clear that human beings began to appear much earlier still, as far back as 100,000 years ago, coexisting in Europe with the Neanderthals for millennia, and supplanting earlier hominids elsewhere (see chapter 7). Perhaps art was being produced more than 35,000 years ago and simply has not survived, but the consensus is that it took the early humans tens of thousands of years to arrive at that level of sophistication.

Why did it take so long for humans to start making representations of the world around them? One view holds that life was so difficult, mere survival such an all-consuming task, that there was no time for the creation of art. In this view, such creativity would have had to wait for the development of more stable, and larger, communities of humans cooperating with one another. In such communities, it is supposed, those with the ability to draw or carve would have been given special status and granted the time to do their wondrous work. To our eyes, much prehistoric art is beautiful. Among people for whom the height of achievement in terms of carving was to make a more efficient flint tool, it must have aroused great awe in some, while others could have regarded it with absolute indifference. Appreciation for this new form of expression clearly grew, however, because it quickly became more and more common.

Another school of thought has emerged regarding the beginnings of prehistoric art. The skulls of early humans dating back to before the first appearance of prehistoric art are congruent with our own cranial structure. Perhaps, however, the brain within that cranium was still less than completely developed—with final connections still to be made that would allow the creation of art. This point of view is somewhat muted because brains are flesh and therefore decay quickly. If, by some remote chance, a fossilized brain of an early human did turn up (as an apparent dinosaur heart recently did), there would be no way to dissect it and determine what differences existed. Even today, the exact workings of the brain remain largely obscure. Nevertheless, it is pointed out that our children start making pictures as soon as they have the dexterity to hold a crayon. It seems innate, and even children who grow up to have little or no artistic talent do it. Was that instinct there 50,000 years ago?

The art created around the world in prehistoric times shows great uniformity in terms of the techniques and mediums used to create it. Charcoal and mineral pigments such as manganese oxides were used for drawings and paintings, while etchings and carvings were produced with stone tools harder than the limestone wall or ivory tusk being worked on. Only animals, human figures, and abstract signs or geometrical figures were depicted. Neither fruits and flowers, nor landscapes, appear. These commonalities do not extend to the symbolism of the art, however, which varies greatly around the world, clearly reflecting an enormous diversity of myths and customs, even in contiguous areas of France, South Africa, or India.

According to many experts, the similar techniques reflect the fact that prehistoric artists used the pigments and tools most readily available the world over. In contrast, the diversity of symbolism makes clear that these minicultures did not interact to the point that widely shared beliefs and customs could take hold. It has been established in recent years that even the last Neanderthals, who lived at the same time as Cro-Magnon humans, were able to learn to improve their tool making by following the example of this new, cleverer species. Nonetheless, the kind of cultural interaction that would eventually give birth to widespread coherent societies in Mesopotamia and Egypt was not yet occurring.

In the 1950s, a prominent theory among anthropologists held that "civilization" arose when small separate groups of human beings encountered one another for the first time. The differences between their customs and myths would create perceptual shock waves that would spur both groups to change for the first time in hundreds, even thousands of years. The "shock of the new" would alter static societies forever, creating inevitable conflicts but also the seeds of future growth. The advent of carbon-dating techniques during the late 1950s poked a great many holes in this theory, as it became clear that such interactions had taken place, affecting such things as tool making, without changing the more symbolic aspects of separate groups. In this respect, it should be noted that decorated caves in neighboring areas of the Dordogne in France have a common roof-shaped ("tectiform") design not used in other regions at the same time. But this shared "architectural" feature does not carry over to what is painted in the caves, which remains very distinct.

Thus, while the creation of symbolic paintings and carvings serves as a clear line of demarcation between *Homo sapiens* and the hominids that preceded us, it does not appear to have acted as a stimulus to what we now call civilization. That would have to await the development of written language and mathematics— and it would be a long wait. The earliest examples of pictorial art go back 30,000 years, but written language would not develop until a mere 7,000 years ago, and mathematics did not appear until 5,000 years ago.

"Civilization" is a tricky word. Most dictionaries start out by defining it as a process—civilizing, or becoming civilized. Second, it is defined as a condition, one that involves social organization of a high order and advances in the arts and sciences. A third meaning has to do with references to an entire culture, whether that of a nation-state, such as Japan, or a period in history, such as the Golden Age of Athens. Probably the most controversial definition is the fourth one, which has caused a great deal of dissension. *Webster's Unabridged* puts it this way: "the countries and peoples considered to have reached a high stage of social and cultural development." This fourth definition demands that a question be asked: "Who says? Who is making the judgment, and is it really all that considered?"

The problem is acutely evident in the conquest of the Americas by white Europeans from the voyage of Columbus in 1492 to the final defeat of Native American tribes in the United States in the nineteenth century. The peoples of the Americas were almost universally regarded as "savages" by Europeans, despite the fact that the Maya (as we show in detail in chapter 13) knew more about astronomy than any European scientist at the time they were conquered. In North America, the Iroquois League, a confederacy of Eastern Native American tribes, would be suggested by Benjamin Franklin, in the 1740s, as a model to be followed by the colonies if they were to form their own government. In the Iroquois League, women had the vote, as they would not under the original United States Constitution, and as they did not in the Athens of Pericles, which is often cited as giving birth to the idea of Western democracy. Who's really civilized here?

To avoid such conundrums, it seems important to hew to a narrow definition of civilization in the following pages. Social mores are not at issue here, nor are religious beliefs, exploitation of other human beings (including women), or forms of governance. It doesn't matter whether an individual had several wives, was a cannibal, or kept slaves. We're not talking about moral issues, period. Instead, we look at the how, why, and when of the great dividing line between the early humans, who were little more than very smart animals, and their descendants, who began to create rudimentary languages and went on to discover basic mathematics. Civilization in this sense started with the creation of words to describe the external world, and of symbols that could be used to keep track of multiple things and to facilitate the trading back and forth of those things.

The evolutionary development of a larynx placed lower in the throat is one of the crucial factors setting humans apart from earlier hominids. Not only did it allow a lower and more modulated vocal pitch, but the change is directly related to spinal-cord developments that allowed an erect posture and a head position that made a larger cranium possible. This linkage of the lower larynx and larger brain may or may not mean that spoken language developed fairly quickly. It is impossible to know when that occurred. Anyone who has traveled in a foreign country with a language he or she does not speak knows that it is possible, when pressed, to convey fairly sophisticated needs through gestures, facial expression, and sounds of protest, entreaty, or delight, which bear only a superficial resemblance to actual language. Studies of chimpanzees over the past few decades have revealed that they can communicate a great deal to one another by such means. Indeed the famous female chimpanzee Washoe was taught to use sign language used by the deaf to make signs for 100 words. The degree to which she fully understood what she was doing remains controversial, but her accomplishments do make clear that early humans might have been able to get a lot across without using oral communication.

How and when spoken language developed is something that cannot be known. Indeed, even the link between spoken language and written language can be unclear. The Romance languages, including Italian, French, Spanish, Portuguese, and even Romanian, are all derived from Latin, which was of course a written language, but the Slavic family of languages (Russian, Polish, and Serbo-Croatian) and the Germanic family (English, German, and Danish) are a different matter. As Merrit Ruhlen of Stanford University puts it, "The usual situation is that the ancestral language was not a written language and the only evidence we have are its modern descendants." Because there are no written records of ancestral languages that gave rise to the Germanic or Slavic languages, "these two languages—which must have existed no less than Latin—are called Proto-Germanic and Proto-Slavic, respectively." Little wonder that the Romans, with their great Latin literature, underestimated the "barbarians" from the north, who had no written language but still managed to sack Rome.

Beowulf, the oldest English epic—written in an Old English so unlike modern English that it must be translated—was not written until sometime early in the eighth century. The fact that English, the most flexible of all languages, with its roots in the Germanic family but a propensity for borrowing from any language on the face of the earth, should have been so late making an appearance can be somewhat deflating (sorry, but no one was writing anything down in King Arthur's Camelot), but it does tell us something important. There was a great deal of English history before there was a written language, and although much of that history is lost to us because of the lack of writing, sites such as Stonehenge make it evident that a lot was going on, and clearly there was a fairly sophisticated spoken language. Still, history depends on writing, and that is why the first civilizations we really know much about are those of the ancient Sumerians and Egyptians.

The Sumerians inhabited the region in the area of the Tigris and Euphrates rivers known as Mesopotamia (literally "between rivers"), where Iraq now holds sway. Mesopotamia has long been called the "cradle of civilization" because it was in this fertile valley that both written language and mathematics are believed to have first arisen. The written records were used to keep track of— what else—tax receipts. They were written in *cuneiform,* a system of writing in which the characters are formed by wedgelike strokes, usually on clay tablets. It was very similar to the better known Egyptian hieroglyphics. The fact that it was inscribed on clay meant that it was preserved for modern archeologists to uncover and linguists to decipher—and that brings up a point that needs to be considered: It is possible that some earlier form of writing existed, but that it was applied to animal skins, which inevitably disintegrated.

This Sumerian clay tablet is a record of loans and payments of barley that were issued to workers at several temples. Dating from 2048 B.C. (the forty-seventh year of the reign of Shulgi, king of Ur in what is now southern Iraq), it is written in fully developed Sumerian cuneiform script. Courtesy the British Museum, Department of Western Asiatic Antiquities; 14318.

The earliest examples of cuneiform writing date to about 5000 B.C. In 1998, a German archeology team found Egyptian hieroglyphics in Abydos, an ancient religious center in the south of Egypt near Luxor. They have been carbon dated to 5300 B.C., challenging the Sumerian primacy. The Egyptian find turned out to be tax records, as well. Thus, regardless of which culture actually was the first to produce a written language, the purpose was the same—keeping track of the money collected from citizens by the ruling classes. Because a great deal of pictorial art that precedes written language is clearly religious in nature—all over the world—there are those who would like to think that human civilization had a spiritual impetus. But here we see the "values" aspect of concepts of civilization creeping in. Religion was highly developed in both the Sumerian and Egyptian cultures, to be sure, but when it comes to actual writing, what we have are tax records. This distinction makes sense: Religious conviction can be measured by devotional habits—you can see someone kneeling or prostrate in prayer. The illiterate can be as religious as the lettered. To keep track of taxes paid, however, you must devise some way to write them down, and it is certainly not incidental that mathematics was also invented in Mesopotamia.

For centuries, it was thought that mathematics had been invented in ancient Greece—Aristotle, Pythagoras, and their cohorts provided eloquent evidence. In 1877, the deciphering of the Egyptian Rhind Papyrus gave indications that something had been happening before the Greeks, but it was not until the 1920s that Mesopotamian clay tables were revealed to antedate any other mathematical source. The Sumerian counting system of the third millennium B.C., like the Egyptian, used an additive decimal system, with a base of 10. The Babylonian empire that replaced the Sumerians in the second millennium B.C. changed to a place-value system, which was far more flexible, even though it had a base of 60—which, though it may sound strange, is divisible by 2 and 5, as well as 3 and 4, unlike base 10. The Babylonian scribes began to work on more advanced concepts, solving both linear and quadratic problems in ways that would not seem strange to anyone who remembers their high school algebra. Some of these problems went beyond any practical needs; in other words, mathematics as a worthy study in and of itself had been established more than a thousand years before the birth of Christ, and centuries before Pythagoras set to work.

Thus, both written language and mathematics had their birth in this "cradle of civilization" in the space of 2,000 years, from 5000 B.C. to 3000 B.C. Yet in other parts of the world, things moved more slowly. The Chinese language did not assume written form until 1400 B.C., and, as we have seen, Germanic and Slavic development was even slower. These differences can be seen as the result of many different factors, from climate to economic systems, and the questions they raise can be (and are) almost endlessly debated by scholars. Although opinions may differ as to why some cultures lagged after others in this regard, there is nothing inherently mysterious about these matters. The time differences are not that extraordinary.

What *is* extraordinary is that it took human beings nearly 100,000 years to reach the point that written language and mathematics became necessary—or is it became *possible*? Once written language and mathematics existed, they spread and developed with extraordinary rapidity in the parts of the world adjacent to their beginnings. The "glory that was Greece" was just around the corner, yielding the plays, poems, philosophies, and scientific theories that inform Western civilization to this day. Obviously this "Big Bang" in terms of human culture was an explosion waiting to happen. As soon as it did, it gathered a force that looks inevitable, but the question still remains why it took so long.

What was going on in the minds of human beings during the first 100,000 years of their time on Earth? There are clues, perhaps, in the caves of Lascaux and the sacred places of Aborigines in the outback of Australia. These people were already capable of making symbolic drawings, some of them very beautiful, even if their meanings are lost to us. It can seem, from our point of view, only a short step from the curious geometric shapes found in prehistoric art to the organized drawings and symbols that the Sumerians and Egyptians turned into language. But obviously it was a very long journey indeed, one that took more than a dozen times as long as the ever-more-rapid one from the first cuneiform writing to the World Wide Web.

It was long believed that evolution takes a great deal of time to do its work, but there are beginning to be questions about that. As we saw in chapter 2, new evidence suggests that the beginnings of life on Earth may have taken much less time than was once thought—and we are evolving still. Suddenly, in the second half of the twentieth century, more and more children around the world began to be born without the unnecessary wisdom teeth that have afflicted humankind from time immemorial. Also, some children also began to be born with one less set of vertebrae in their backs—which will mean fewer backaches for a species that walks upright instead of on all fours. Why did that take so long, and why did it start happening so suddenly? Such mysteries of the evolutionary process raise a larger speculative question. Were early humans still not quite fully formed in terms of the billions of tiny connections that exist within the human brain? Were there linkages yet to be made that would allow us to write down our histories and our visions of a better future, connections that would make it possible for us to begin to calculate our way toward the stars?

We do not know—and it is unlikely that we ever will.

✳ To investigate further

Hooker, J. T. Introduction to *Reading the Past: Ancient Writing from Cuneiform to the Alphabet* (by Latissa Bonfante et al). New York: Barnes & Noble Books, 1998. Written by six specialists, this book traces the development of writing from pictograms to the modern alphabet, while Hooker's introduction gives a wider-ranging overview than the individual essays.

Ifrah, Georges, and David Bellos: *The Universal History of Numbers: From Prehistory to the Invention of the Computer*. New York: John Wiley & Sons, 1999. This book has received rave reviews in several countries for its scope, depth, and readability. It attempts to present the history of the human race through its relationship to numbers—and succeeds.

Potter, Simeon. *Our Language*. New York: Penguin Books, 1976. There are numerous books on the development of the English language. This one, despite being published 25 years ago, remains one of the classics in the field.

Note: This chapter was based on a great many different sources. The preceding books therefore focus on several quite different aspects of the overall subject.

Chapter 9

How Do We Learn Language?

I have a good friend of many years who rose through the ranks to become one of the most important simultaneous translators at the United Nations. Maria has a truly remarkable gift for languages, even among her peers. I have been at parties she gave that included among the guests people of several different nationalities, some of them as comfortable with the English language as she is, but some of them not. She was perfectly capable of telling an anecdote, using several different languages, in a way that made it possible for every person in the room to follow the story. Surprisingly, she didn't tell a part of the story in one language, and then translate it into two or three others as she went along. Instead, she would tell two sentences of the story in English, another two in French, another couple in Russian, never repeating any facts in a different language, but somehow keeping all of her listeners on track no matter what language they spoke.

Even someone as gifted as Maria does not remember how she learned her three basic languages. Her mother was Spanish, her father Italian, and she grew up in a French household in Paris. All three of those languages are second nature to her. English she learned at school, including a year spent in England. Languages that she went on to learn as an adult, German and Russian in particular, were, as she puts it, "a lot of bloody hard work."

It is precisely because we learn to use languages—just one in my case, three in Maria's—at too early an age to remember the process that makes the subject fascinating, frustrating, and fraught with disagreement. The entire field of what is now called *psycholinguistics* is caught in a terrible bind: A deep mystery exists about how language "happens," but the only true witnesses are too young to explain to us what is going on, and by the time they are old enough to begin to convey what they experienced, it is utterly forgotten.

For most of human history, the mystery of language acquisition was in fact largely ignored. Children learned to speak the language of their parents, as might be expected, and that was that. If the child learned to speak while the family was living in a foreign country, it often happened that he or she would also quite effortlessly pick up that second language as well, even if the parents had some difficulty in speaking it. The very young children of immigrants to the United States, for example, seldom have any difficulty switching back and forth between English and the language their parents speak at home—although if the child is nearly grown by the time the family comes to America, that teenager may have to work quite hard to learn English. In fact, the older a child is when he or she is exposed to a second language, the more work it is likely to take to learn the new language and the greater the chances are that a foreign accent will persist throughout life.

Exactly because all normal children begin to speak the language of their parents with relative ease, this remarkable transformation was largely taken for granted until very recent times. Many saw it as just one more example of God's gifts to the human race. Even those holding more secular views, however, had little interest in how children learned language. It was something that happened, and eventually it was hoped that they might learn enough to be worth listening to. "Children should be seen and not heard" is a saying that reflects far more than notions about manners. People simply didn't much care what was going on in children's heads—or at least, men didn't, and it was men, of course, who ruled the intellectual roost. In the nineteenth century, and well into the twentieth, it was mothers who were expected to teach their children to use language properly. When the child had progressed far enough, the fathers would start to take notice and guide the child in more "elevated" matters. The mother would speak to the child in a way that changed with his or her age, starting with "baby talk" and ultimately correcting grammar and word usage almost absentmindedly. Ironically, as women have gained their own place in the academic and scientific world, such casual dialogues between mother and child (now termed "parentese") have become a fiercely debated subject, as we see in this chapter.

Despite women's primary role in child rearing, it was a man who forced the world to take a new look at what was going on in the heads of children. Sigmund Freud, with his theories about the workings of the subconscious, and his emphasis on how repressed childhood experiences resulted in adult neuroses, elevated the child to a new position of importance. If Freud was correct, then a great deal must be happening inside those seemingly innocent little heads. As scientists began to pay more attention to what children were actually doing and apparently thinking, linguists began to take a real interest in how living children learned to use language. The primary focus of linguists had been on the historical development of the world's languages, and on decoding ancient Sumerian, Egyptian, and Mayan tablets, unlocking the secrets of dead civilizations. With the new emphasis on living languages and their acquisition, scholars began to recognize that the mystery of how children learn new words and then start putting them together in not just understandable but grammatical ways was a very deep one indeed.

The foundations of psycholinguistics were laid down by Swiss psychologist Jean Piaget, starting in the 1920s. Piaget developed a theory of *cognitive development* (how human beings synthesize and make use of the information their senses record), which held that such development proceeds in genetically determined sequential stages. Not only did he believe that the way children learn language is innate, but also that each new step forward was taken in a precise order, determined by the maturation of the child. Born in 1896, Piaget lived until 1980, long enough to see his theories challenged from several different directions. But even those who

Swiss developmentalist Jean Piaget revolutionized the study of child development and language acquisition beginning in the 1920s by observing, testing, and often playing with selected children at successive ages in an interactive way that was entirely new. His techniques laid the groundwork for modern studies in cognitive development. Courtesy Wayne Behling, Ypsilanti Press, Michigan.

believe that he was wrong about many things will admit that they owe him a great deal, primarily because of the way he went about obtaining the information on which he based his theories. Piaget spent more time talking with children than anyone ever had before, spending a great deal of his life sitting on the floor playing with them, asking them questions, and giving them problems to solve. That approach to gathering facts—or seeming ones—about how children think and learn became one of the basic tools of cognitive research, and it remains one today.

But as more and more psychologists and linguists began studying children in this manner, they kept finding evidence that conflicted with many of Piaget's theoretical conclusions, particularly with respect to his rigid, age-based stages of development, which seemed to show that certain problems with solutions that required logic were beyond the capacity of 7-year-olds, but easy enough for 13-year-olds. As Morton Hunt noted in his best-selling

1982 book *The Universe Within,* researchers who repeated Piaget's experiments did not always get the same results. "This may mean that Piaget's findings do not have universal validity. Perhaps the children Piaget worked with were a special, privileged group; perhaps the way Piaget and his coworkers asked questions of the children elicited reasoning the children might not have done spontaneously; or perhaps Piaget, who by training was strongly disposed to favor logic, overinterpreted the children's answers." The problem is that many psychologists believe that human beings do not use formal logic much of the time, but can draw on it when they have to. As Hunt suggested, Piaget's emphasis on logical problems could easily create a false impression: Just because you see someone swimming doesn't mean that swimming is the person's usual mode of getting around. There is the additional problem of qualitative differences. A kid doing laps in the pool is not necessarily going to win an Olympic Gold Medal in the 400-meter freestyle swimming contest.

In fact, the kinds of tests that Piaget used (such as vials of colorless liquid that can be affected, or not, by adding a coloring agent) are widely under attack these days as a measure of intelligence. I can personally attest to the vagaries of such results. When I was in high school, I got very high grades in English and history but very poor ones in algebra and geometry. Specifically, I had problems with the math. In geometry I had a very strong grasp of the spatial elements involved, and because there were a number of such problems on the SAT (a pile of cubes was shown, for example, and test-takers were asked to determine the number of sides that were not visible) I did quite decently, to the surprise, even the annoyance, of some of my teachers. We all have mental strengths and weaknesses and, however much many politicians and educators hate to admit it, tests do not always measure them well. For several years, my father, a teacher of American history, went to Princeton, New Jersey, to take part in putting together the advanced-placement tests in that subject. His chief concern was getting rid of questions that the average student would have no trouble answering, but that the very best students might well fumble, understanding that there were multiple, or more complex, answers. Such differences in how different people think, and the different levels of knowledge they have achieved in various areas, are not just the bane of standardized testing, but also a confounding factor in the kinds of experiments that psychologists and linguists use to probe the mystery of how we learn in the first place.

Since the 1960s, psycholinguistic researchers, recognizing some of the problems inherent in working exclusively with children at various ages, have turned to a number of alternative approaches. Some have studied individuals whose ability to talk has been affected by strokes. Others have worked with adults having mental deficiencies. Also, famously, thanks to the mass media, some have worked with chimpanzees, trying to teach them to use sign language to communicate in "human" ways. Books on psycholinguistics often give the results of such studies in great detail, using them as evidence to support one theory or another. While many of these experiments are very clever, they often come across as more anecdotal than scientific—not unlike this chapter.

Reading about these experiments, I tend to think back to a family I knew when I was growing up. The father was a distinguished teacher of foreign languages, and the mother was the daughter of an important American diplomat of the 1930s and 1940s. They had a son who, to their great dismay, did not start to talk at two, or at three, or at four years of age. He was given every test imaginable, of course, and there did not seem to be anything wrong with him physically. What's more, in every other way he behaved like a normal child. Finally, when he was five years old, he started talking—a blue streak, and with a very sophisticated vocabulary for his age. His parents were overjoyed, but many of the experts who had examined him and worked with him quickly found themselves infuriated. When they asked him why he hadn't talked before, he had a very simple answer: "I didn't want to." He went on to become a top student, but the anger of the experts can certainly be understood—this is the kind of case that can knock any number of theories, based on numerous experiments, into a cocked hat.

Indeed, one of the chief activities among psycholinguists is pulling the rug out from under the experiments of other researchers and theorists. It is all too easy to do, which may be one of the reasons why the work of Massachusetts Institute of Technology (MIT) linguist Noam Chomsky came to dominate the field in the 1960s and 1970s. Steven Pinker, Chomsky's successor as the star of psycholinguistics, tells (in his 1994 book *The Language Instinct*) of a couple with a severely retarded child who was nevertheless a very loquacious and imaginative conversationalist. Having read about Chomsky in a magazine, the girl's parents wrote to him suggesting that he might be interested in studying her. Pinker wryly but affectionately commented, "Chomsky is a paper-and-pencil theorist who wouldn't know Jabba the Hutt from the Cookie Monster," explaining why Chomsky thus steered the girl's parents toward a researcher who worked directly with children. This "ivory-tower" position effectively protected Chomsky from the internecine warfare that was waged among field researchers, a fact that no doubt has helped him to maintain his preeminence in the field. Although many others put his theories to the test in experiments, he was not directly involved.

Chomsky also has a brilliant mind. Although he was already well-known in the field at age 31, he came to prominence in 1959 with a scathing review of a new book by B. F. Skinner, the high priest of behavioral psychology. Skinner was himself very famous for his theories concerning the malleability of human behavior: With the right techniques, he asserted, you could change human behavior to any model you wished. He was also notorious for developing the "Skinner Box," any of several cages used for animal experimentation, and the "Baby Tender," a glass-enclosed crib in which his younger daughter slept at times during the first two years of her life.

Skinner came to Chomsky's attention by writing a book called *Verbal Behavior,* in which he claimed that language was nothing more than a "habit" established by conditioning. Chomsky responded, in effect, "utter nonsense," pointing out that children constantly create new sentences that are unlike anything they have ever heard, something that cannot result from the kind of imitation presumed by "conditioning." Chomsky accused Skinner of "play-acting" at science, a charge from which Skinner never really recovered. I myself took Skinner's introductory course at Harvard, which I found very interesting, but perhaps not in the ways intended. We were assigned his novel, *Walden Two,* about a utopia in which everyone was conditioned to play a particular role and be happy doing it. I was fascinated at the very clever trick he kept playing on the reader. He would leave a gaping hole in the logic of the story, and just when you were ready to throw the book across the room, he'd plug that hole—which had the effect of putting you thoroughly off-guard, so that he could slide quite a lot of dubious material past you in the next few pages. This is not to suggest that people cannot be "conditioned," but rather that it is clearly a far more complex process than Skinner and his followers believed. The confrontation between Chomsky and Skinner has haunted the behaviorist school of thought in terms of language acquisition ever since.

Behaviorism is the apogee of the idea that "nurture" (the lessons that children are given by parents and other authority figures such as teachers and clergy) is far more important than "nature" (the biological human animal, including the various genetic constituents of individual people). The nature-nurture argument is as old as human history, and leaving science entirely out of it, one or the other has always been in ascendancy or decline according to the prevailing political ideology of the moment (for example, take a look at arguments about prison reform). The very fact that the nature-nurture debate is so easily bent to political purposes—and can so easily become tangled up in religious views—makes even scientific theories about language acquisition highly vulnerable to outside influences. Despite these potential pitfalls, Noam Chomsky was such a brilliant thinker that his ideas seemed for a time to be fairly impervious to these problems, even though he came down strongly on the side of nature. The "language faculty," as he termed it, was a genetically determined brain structure with a "preexistent knowledge" of how "the things and actions represented by noun phrases and verb phrases are related to each other as agent, action and object," as Morton Hunt puts it in *The Universe Within*. One of Chomsky's examples uses two sentences with the same structure—at least on the surface:

"John is easy to please."
"John is eager to please."

Try changing these sentences around:

"It is easy to please John."
"It is eager to please John."

Innumerable studies have shown that children recognize the first of this second pair as making sense and realize that the second is wrong unless "it" is an enthusiastic pet. Children grasp the kind of difference that exists here, between the surface structure and the deep structure of language, with respect to hundreds of similar examples in whatever language they are first exposed to. German has a very different word order than English does (it can seem backward to an English-speaking adult trying to learn German), but young children seem to grasp the rules governing noun and verb phrases in whatever language their parents speak.

Chomsky did not claim, however, that language is innate. To say that is to suggest that language exists in the human brain even if a child is not exposed to it. If it were innate, then even a "wild child" like the famous Kasper Hauser, or abused children kept locked in cellars for years without human contact, ought to have developed languages of their own, even without exposure to any. They do not, although they can still be taught, to some degree. Since the early 1990s, however, many psycholinguists, led by Steven Pinker, have come to the conclusion that language is an "instinct" in human beings, of the same sort that causes spiders to spin webs. "Web spinning," Pinker has written, "was not invented by some unsung spider genius and does not depend on having had the right education or on having an aptitude for architecture or the construction trades. Rather, spiders spin spider webs because they have spider brains, which give them the urge to spin and the competence to succeed." He goes on to acknowledge that this view contradicts conventional wisdom, which holds that language is a cultural invention. Pinker insists that it is "no more a cultural invention than upright posture." Like bats with their sonar or birds that can migrate thousands of miles, we are just one more part of nature's great talent show with our own specialized act: language.

Needless to say, because this concept of language does fly in the face of conventional wisdom, Pinker has his opponents. A great many do not like the idea of a language instinct because it tends to knock the pins out from under some basic concepts of what is good and true in human beings. "Parentese"—the give-and-take between parent and child that starts with baby talk, continues with vocabulary lessons ("see the doggie"), and proceeds to the occasional grammar correction—is seen by most people as a "warm and fuzzy" aspect of parenthood. When you tell people that children are not learning language from their parents any more than they learn to walk by emulating their parents, you are striking at the heart of that long-held "apple pie" kind of belief. While it is clear that to activate children's language instinct, children need to be around people who talk, innumerable studies of many different kinds, including anthropological ones that take note of societies in which parents do not say much to their young children, show language acquisition to be quite detached from whether children are addressed directly or not. They appear to learn just as quickly when most of the conversation they hear is strictly between adults. In other words, they do it largely on their own, and direct parental guidance counts for surprisingly little.

To some people, this concept is very disturbing. Such cultural antipathy toward the idea of a language instinct is ultimately less of a problem, however, than is an unanswered question: Where in the brain, exactly, is this language instinct located? One thing that is relatively clear is that it has to be located in the left hemisphere of the brain. That fact was established back in the 1860s by French physician Paul Broca, who dissected the brains of several patients who had severe language problems. In one after another, he found lesions in the left hemisphere of the brain, affecting what is called the perisylvian cortex, which surrounds the cleft between the temporal lobe and the rest of the brain. Still, exactly where the language "controls" reside in this area is a mystery. How they work is an even deeper problem.

There is a great deal of important new work being carried out in the field of neuroscience. For example, positron emission tomography (PET) makes visible the chemical functioning of organs, including the brain, moving beyond the structural images produced by X-rays and magnetic resonance imaging (MRI). New types of brain cells, called neurons, are constantly being discovered. But what is being revealed is greater and greater complexity, and that presents problems of the kind that have bedeviled quantum physics, as we will see in chapter 16. Great numbers of neurons, like great numbers of subatomic particles, make it even more difficult to establish the relationships between them. As John Horgan reports in *The Undiscovered Mind,* 1981 Nobel Prize winner Tortsen Weisel even objected to the 1990s being declared the "Decade of the Brain," saying that comprehending how the brain works will require "at least a century, maybe even a millennium."

As of now, even what is known about Broca's area is largely a matter of inference. Speculating that the brain has specific regions to deal with nouns, others to deal with verbs, Pinker writes, "Perhaps the regions look like little polka dots or blobs or stripes scattered around the general language areas of the brain. They might be irregularly shaped squiggles, like gerrymandered political districts. In different people, the regions might be pulled and stretched onto different bulges and folds of the brain." Such patterns do appear in better-known systems such as that controlling vision, but we simply do not know enough to do more than speculate.

In the twentieth century, the study of language acquisition at least became a genuine field of scientific endeavor, but theories, inferences, and speculations do not add up to facts. Seeming facts, like those developed by testing children at various ages, are always open to challenge. And the degree of anecdotal material— observations of people with language problems—are all too similar to eyewitness reports in the news or criminal trials: the more observers, the greater the variations in storyline. It has been said that we know more about the behavior of subatomic particles than we do about the human brain, and that seems likely to remain the case for a long time to come. Because we are the only creatures on Earth that talk, the mystery of language acquisition may be the last aspect of brain function we will come to fully understand.

⚛ To investigate further

Hunt, Morton. *The Universe Within*. New York: Simon & Schuster, 1982. Although there has been a great deal of ferment in the field of psycholinguistics since this book was written, it remains one of the most accessible books on the subject and covers a lot of important material about the development of this new science that more recent books simply don't have room to deal with.

Pinker, Steven. *The Language Instinct*. New York: Morrow, 1994. In this book Pinker gives the clearest and most complete account of his theories about the existence of a language instinct, and how his own work relates to that of others. He is a lively writer, as well, drawing cleverly on pop culture to make various points, and displaying a fine sense of humor.

Pinker, Steven. *Words and Rules*. New York: Basic Books, 1999. Pinker's follow-up to *The Language Instinct* further develops his theories and shows how they relate to a wide variety of subjects, from the peculiarities of the English language to the history of Western philosophy. This is in some ways a more difficult book than its predecessor, but readers with a fondness for word games may find it more appealing.

Chomsky, Noam. *Reflections on Language*. New York: Pantheon Books, 1975. Chomsky remains an important and influential figure, and this book, published at the height of his early success, is a classic of its kind. It consists of a series of lectures augmented by an additional long essay. Although it requires very close reading and is quite academic in its approach, dedicated students of language will find it highly rewarding.

Horgan, John. *The Undiscovered Mind*. New York: Free Press, 1999. Horgan does a fine job of explaining the latest techniques used to investigate how the human brain works, and the theories they have given rise to, but concludes that vast difficulties remain. This book is a good antidote to much of the hype on this subject at the start of the twenty-first century.

Chapter **10**

Are Dolphins As Smart As We Are?

Human beings have always been fascinated by dolphins. Herodotus, whose account of the Greek wars with the Persians was the first secular narrative history, told a story, picked up and used by Shakespeare in *Twelfth Night,* about the poet Arion's attempted suicide when his ship was attacked by pirates. Sensing that all was lost, he sang a final keening song of farewell and leaped into the sea to drown himself—but was rescued by a dolphin that carried him several miles to shore. To this day, stories crop up occasionally around the world about sailors and fishers who have supposedly been rescued by dolphins. Four centuries later, Plutarch, another of the great formative influences on Western literature, wrote one of his famous moral essays on "Whether Land or Sea Animals Are Cleverer," and had this to say on the subject of dolphins: "to the dolphin alone, beyond all others, nature has granted what true philosophers seek: friendship for no advantage."

The playfulness of dolphins as they swim alongside ocean-going vessels, the varied calls they make to one another, sometimes sounding as raucous as those heard in a high-school cafeteria, at other moments haunting and plaintive, and the perpetual grin they seem to wear, have endeared them to several thousand generations of humans. It has often seemed even to the casual listener that they are speaking a language of their own. It was not until the second half of the twentieth century, however, that dolphin study became a full-fledged scientific endeavor. Anthropologist Gregory Bateson, a colleague of Margaret Mead in her renowned work in New Guinea (they married in 1936), was so taken with dolphins that he began to study their behavior, and in 1965, he was able to show that they lived in close-knit social groups with a clear leader, much as primates do. At the same time John C. Lilly, the noted researcher on expanded states of consciousness in humans, was working with dolphins to try to determine the extent of their communication abilities, both among themselves and in relation to human beings.

Bateson's findings have been corroborated again and again. There is no question that dolphins form complex social groups. Lilly's work, although many other researchers have been inspired by it, has always been controversial. One of the tests he carried out shows why his research was regarded as exciting by some and fatally flawed by others. Lilly decided to see whether he could teach a dolphin (called Number 8—he avoided using names) to repeat a whistle of a particular pitch, duration, and intensity. The dolphin was rewarded with food when he got it right. As is often the case with such tests, the dolphin caught on very quickly to the rules of this "game." Then Number 8 seemed to change the rules of his own accord, raising the pitch of each subsequent whistle emitted through his blowhole. Next, Lilly noticed that although the blowhole was moving as though a sound were being made, Lilly could hear nothing. It was clear what had happened. Dolphins are perfectly capable of making sounds, which can be monitored electronically, that are beyond the range of human hearing. Lilly was delighted that the dolphin was restructuring the game— it was another sign of the adaptive intelligence of which he believed them capable. Nonetheless, rules had been established, and because Lilly could not hear this sound, he offered no reward. The dolphin, judging by the twitching of its blowhole, tried twice more to reap the reward with sounds that could not be heard and then started making whistles that Lilly could hear once again.

To Lilly, the dolphin's behavior was evidence of high intelligence, an ability to test its teacher that was remarkable in itself, and, even more impressive, to grasp the problem that these high-pitched sounds were causing and remedy it. To Lilly's critics, however, absolutely nothing had been proved. So a dolphin was able to mimic a few whistles. Maybe, they countered, the fact that the dolphin changed the rules of the game was a matter of being stupid instead of clever. Dolphins might be playful, but ascribing intent to such fooling around seemed a stretch to the doubters. The implication that the dolphin was deliberately raising the pitch was just Lilly's interpretation. It could have been accidental, or worse, it could have been caused by an inability to focus on the matter at hand—getting food by producing a particular whistle. In other words, the supposed intelligence was in Lilly's head rather than the dolphin's. This kind of criticism was leveled against many other experiments Lilly carried out—as it has been against a lot of experiments with chimpanzees by other researchers. Lilly himself had also worked with chimpanzees and in many cases found that dolphins could learn to push the right button (literally in some experiments) with vastly fewer tries than chimps could. To the critics, however, that finding was a matter of comparing apples and oranges. Maybe some tests were simply better suited to dolphin behavior.

Quite a few scientists simply don't like tests for animal intelligence in any way, shape, or form. They believe there is a tendency to *anthropomorphism*—falsely attributing human traits to animals—on the part of such researchers, which inevitably skews results. Lilly had an additional problem, however, in terms of garnering scientific respect. He was also interested in extrasensory perception in human beings. Still later, he took a deep interest in Carl Sagan's work with Search for Extra-terrestrial Intelligence (SETI), the search for radio signals emitted by alien civilizations in space. In fact, he had even suggested that we would be wise to learn to communicate with dolphins because that would help give us a grounding for dealing with alien intelligences in the future. That kind of statement truly sets some scientists' teeth on edge. Nor did it help that Lilly's work inspired the best-selling novel *The Day of the Dolphin,* a richly rewarding book by Robert Merle later made into a dreadful 1973 movie by Mike Nichols. The dolphins were the heroes of the book, demonstrating extraordinary abilities not only to carry out specific tasks, but also to tell right from wrong—anthropomorphism run amok!

Nevertheless, other researchers have continued to work with dolphins, and some have had results that successfully support Lilly's high regard for the intelligence of dolphins. An experiment carried out by Javis Bastian with two dolphins named Buzz and Doris demonstrated an ability to communicate what we would consider abstract ideas. The dolphins were placed in a divided pool that allowed them to see one another on the opposite sides of a clear barrier. Double switches and signal lights were installed on both sides of the pool. If the signal light emitted a steady beam, the dolphins were supposed to push the right switch. If the light blinked, they were to push the left switch. They learned to do this with no problem, getting a food reward after carrying out the test correctly.

Then the conditions were made more difficult. Buzz was required to push the correct button first, while Doris held back. Then she had to press the same button to get the reward for both of them. Once they had mastered that, a wall was built in the pool so that they could no longer see one another, and only the signal light on Doris's side was activated. The two dolphins could still hear one another, however. When the steady light came on, Doris waited for Buzz to push his button first, as they had been taught in the second step of the experiment. Of course, because Buzz's light had not been activated at all, nothing happened. Doris then made a sound. Buzz immediately pressed the right-hand button on his side—even though no light was on that he could see. Doris then took her turn, and they both received their fish. The test was repeated 50 times, and Buzz usually hit the correct switch, although he made occasional mistakes. Three things were demonstrated in this experiment: (1) Dolphins were able to distinguish right from left—an abstract idea—without problem; (2) Doris was able to communicate to Buzz, through sound alone, whether he should push the right or the left button; and (3) Doris showed problem-solving ability when she recognized that a new situation existed.

Over the years, such experiments, as well as observations of dolphins in their own habitats, have been startling enough to raise real questions about how close to human beings they are in intelligence. John Lilly's early experiments may not have been as rigorously designed as they could have been, but subsequent research of the kind done with Buzz and Doris has supported his high estimation of dolphin abilities. They are very smart indeed— few scientists argue with that anymore. How do they compare with us, though?

One classic method of calculating the probable intelligence of different species is to compare brain weights *in relation to overall body weight*. The bottle-nosed dolphin, the kind we are most familiar with and most likely to encounter, has a ratio of brain weight to body weight that is second only to human beings. On average, human beings have a 2.10% ratio, dolphins a 1.17% ratio. Chimpanzees, which come in third, have a 0.70% ratio. If you look at brain weight alone among these three species, disregarding body weight for the moment, dolphins come in first with an average brain weight of 3.50 pounds (1.75 kilograms). Human brains average 3.00 pounds (1.4 kilograms) in weight, and those of chimps, 0.75 pounds (0.4 kilograms). These are, remember, average figures. Some dolphins have brains weighing up to 5 pounds (2.3 kilograms), but they are also bigger in terms of body size. While these figures are interesting and often used to suggest that dolphins are second only to human beings in intelligence if you stick with the brain-weight/body-weight ratio, and perhaps more intelligent if you consider the brain-weight differential between the two, there are serious problems with such comparisons.

Chris McGowan, a Canadian who is both a professor of zoology and a curator of vertebrate paleontology, thoroughly demolishes the significance of the brain-weight/body-weight ratio in his 1994 book *Diatoms to Dinosaurs: The Size and Scale of Living Things*. He points out the problems by using both simple and complex examples. "A cat's brain, for example, accounts for 1.6 percent of its body weight, while a lion's is only about 0.13 percent, even though the lion is not intellectually inferior." The reason for the differential in this case has to do with metabolic body rates, but while that holds up in many cases, it is far from universal. McGowan discusses a number of sophisticated attempts to relate brain and body size, including that of the early expert in the field, Harry Jerison, who developed a logarithmic graph for nearly 200 species of vertebrates, including mammals, birds, fishes, amphibians, and reptiles. The meaning of the results tended to break down between one group of animals and another, McGowan notes. The difference in brain size is, for example, much greater between large and small individuals among the primates than it is among cetaceans (whales and dolphins).

Even among cetaceans, there are differentials that create questions. The blue whale, for example, is twice the length of a sperm whale, but sperm whales have the heaviest brains that have probably ever existed on the planet, the record being a 49-footer (15 meters) with a 20-pound (9-kilogram) brain, which was killed in 1949. Blue whales, however, are of the *mysticete* (baleen) group, which must consume enormous amounts of small crustaceans to survive, and which have mouths equal to a third of their body length. Those huge mouths are filled with a kind of enormous strainer made of baleen (whalebone), and this eating machinery takes up so much of the head that there is not all that much room left for a brain. Sperm whales belong to the *odontocete* (toothed-whale) group, as do dolphins, and they need larger brains for two reasons: (1) They *echolocate* (using sound to guide them), and (2) they belong to complex social groups, both of which require greater intelligence than plowing through the seas like a giant vacuum cleaner, as the blue whale does.

A further difficulty lies in the functions a brain, of whatever size, must carry out. We know even less about how the sections of a dolphin brain work than we do about the human brain—although there are some similarities in structure—but it seems likely that a good deal of brain space must be given over to the process of echolocating. It is, after all, a sonar system so finely tuned that the U.S. Navy has spent a great deal of money on dolphin research in order to facilitate numerous underwater operations. In addition, dolphins control each breath they take and can concentrate blood in specific parts of their bodies when they dive. If humans could do this, we would be able to consciously overcome asthma or regulate blood pressure. In this respect, dolphins are less instinctive and more in charge of their body systems than we are. That can be looked at two ways. It can be seen as a mark of great intelligence, far surpassing anything we are capable of, or it can suggest, once again, that so much of the large dolphin brain is given over to this kind of regulatory activity that there is less room for abstract thought and language creation.

Language creation—here, perhaps, we get to the crux of the mystery. There is no longer any question that dolphins can communicate with one another in startling ways. They have been observed having what can only be termed conferences under certain circumstances. For example, when groups of dolphins have approached passages that have been strung with underwater microphones on poles driven into the seabed, they have stopped swimming forward while one dolphin went ahead to check out the situation. When the "scout" returned to the group, the dolphins uttered a wide variety of sounds and then made their way forward again as a group. Underwater observers of such behavior—and it has been seen a number of times—have been mesmerized by the seeming discussions taking place.

A report published in *Science* in August 2000 goes further. Vincent M. Janik, a Scottish biologist, analyzed more than 1,700 whistle signals that bottle-nosed dolphins exchanged while swimming along the Moray Firth coast of Scotland. The dolphins routinely responded to each other with identical signals within seconds. Noting that matching communication signals "has been hypothesized to have been an important step in the evolution of human language," he suggests that dolphins are capable of "vocal learning," which is a prerequisite for the evolution of spoken languages. Earlier research by others had made it clear that young dolphins adopt a signature whistle pattern, a form of self-identification that could be interpreted as a name. That specificity would make it possible for one dolphin to send a whistle message to another specific dolphin swimming some distance away.

Doubters protest that such whistles don't reveal sufficient variety to qualify as a language. Even though it may not seem like a

These Atlantic spotted dolphins (Stenella frontalis), photographed in the Bahamas, seem here to possess some kind of intelligence, as one group in tight formation, virtually still, watches a smaller group squabbling among themselves. Photograph by Phillip Colla, all rights reserved.

language to us, however, it might be to dolphins. It is worth considering one of the great code triumphs of World War II in this regard. The Marines recruited several dozen American Navajos to serve as "code-talkers" in the Pacific. They were assigned to different Marine companies, and when messages had to be radioed back and forth, the Navajos would staff the radios, having been trained to use particular words for specific military actions. Because the Navajo alphabet had only recently been set down, the Japanese were utterly stymied. They broke just about every other code we had, but not that one. Thus, it may be that dolphins do have a language—but one that we have no idea how to interpret. In this kind of context, John Lilly's suggestion that learning how to communicate with dolphins might one day help us to communicate with extraterrestrials doesn't sound as silly as many originally found it.

Dolphins are not extraterrestrials themselves. They inhabit Earth, and they have brains that are comparable in size and structure to our own. What's more, they evolved millions of years before we did. We're newcomers. For some reason they seem to like us. They are not always friendly—in recent years, evidence has accumulated to suggest that they can be brutal toward one another at times, and possibly even dangerous to humans in rare cases. Throughout human history, however, they have established a reputation as being not only friendly but sometimes extremely helpful to humans. No one knows why they seem to like us so much most of the time. It has been suggested, not entirely in jest, that they are always grinning because they find us ridiculous. After all, they have been around a lot longer than we have.

Will we ever crack the "dolphin code"? Of course there may not even be one, but a lot of scientists think we need to keep working at the problem. We've taught them to throw switches to get some fish, but the greater question remains, What could they teach us?

⚛ **To investigate further**

McGowan, Chris. *Dinosaurs to Diatoms*. Washington, DC: Island Press/Shearwater Books, 1994. Subtitled "The Size and Scale of Living Things," this is a delightful, wide-ranging book with a personal style. The author, a professor of zoology as well as curator of vertebrate paleontology at Canada's Royal Ontario Museum, connects up a great deal of material in a thought-provoking way.

Carwardine, Mark. *Whales, Dolphins, and Porpoises*. New York: Time-Life, 1998. This Nature Company Guide is a lavishly illustrated tribute to these marine animals.

Lilly, John C. *Communication between Man and Dolphin: The Possibilities of Talking with Other Species*. New York: Crown, 1978. Controversial though Lilly is, he started the debate on this subject, and this book is fascinating reading. Although out of print, it is in many libraries and available on the Internet at sites such as BookFinder.com and alibris.com.

Cousteau, Jacques Yves, and Phillipe E. Diol. *Dolphins*. Garden City, NY: Doubleday, 1975. Cousteau is always worth reading on any marine subject. It is available from used bookstores and through BookFinder.com and alibris.com.

Chapter **11**

How Do Birds Migrate?

They travel in small groups or vast flocks that can darken the sky. At the so-called "staging posts" where they pause for rest stops during their great migrations, on the shores of the Delaware Bay or the Caspian Sea, 100,000 birds of a single species will congregate along a few hundred yards of shoreline. When they reach their nesting areas in the spring, the great flocks disperse, each mating couple choosing a particular nesting spot. We look out the window, watch the nest being constructed in the apple tree, and think, "Can those really be the same birds that chose that tree last year?" We are amazed to contemplate how far they have traveled and that they can find their way across continents and oceans with such seeming ease. How do they do it?

The astonishment that bird migrations arouse in human beings can be seen in Egyptian reliefs that were carved as early as 2000 B.C. Despite millennia of these observations, it was a long time before we began to try to understand, in any scientific way, the how and why of bird migrations. One of the first to write about the subject was Greek philosopher Aristotle, in the fourth century B.C., and he got it all wrong. He correctly identified a few species as being migratory but confused the issue completely by deciding that in the course of their travels, these birds changed from one kind into another. The concept of transmutation, by which a Robin became a Redstart and then changed back again, was widely accepted and repeated until the sixteenth century—proving only that if your reputation is great enough even your worst errors can have a very long shelf life. We can guess, however, how Aristotle became confused: Robins summer in northern Europe and winter in Greece, while Redstarts summer in Greece and winter in sub-Saharan Africa. Their size and coloring are sufficiently similar for Aristotle to assume they must be the same bird in different guises, an avian variation on the transmutation of the caterpillar into the butterfly.

By the sixteenth century, as explorers sailed the globe and the Americas were settled by Europeans, it became clear that such fanciful ideas were wrong. But a new argument arose. Naturalists were convinced that these same birds were, however miraculously, commuting vast distances, in some cases from one continent to another. Because the naturalists could not begin to explain how tiny songbirds weighing only a few ounces could manage to traverse distances that humans themselves were only beginning to conquer, other theorists came up with an entirely different idea: Birds did not migrate at all, they insisted. Instead, they disappeared from a given locale for part of the year because they hibernated. If a creature as large as a bear could hibernate, surely small birds could easily do the same. The backers of this theory had their own problems of proof, though: If the birds were hibernating, where were they doing it, and why could no one find their winter lairs?

Hibernation, it was eventually discovered, does occur among birds but is highly unusual—the Nuttle's Poorwill of the California deserts region is a rare example. There are other birds, particularly among the owl family, that neither hibernate nor migrate. Barred Owls and Great Horned Owls, for example, are able to sustain themselves in one location year-round. The smallest of all owls, the 6-inch (15.2-centimeter) Elf Owl of the American Southwest, does migrate to Mexico, however, because it feeds on insects rather than small mammals, and the food supply disappears during the winter months.

It is lack of food supply, rather than cold weather per se, that lies behind bird migration. Unlike human "snowbirds," who winter in the warmth of Florida and return north for the summer, actual birds are not in search of more comfortable climes but of basic sustenance. That search can take them on journeys that we find staggering even in the age of jumbo jets. Arctic Terns migrate from nesting grounds on the Arctic Circle down the coasts of Europe and Africa to the Antarctic every year. Bobolinks fly 5,000 miles (8045 kilometers) from Canada to the grasslands of Southern Brazil, Argentina, and Uruguay. Some species reach incredible heights on their travels, with Bar-Headed Geese achieving an altitude of 29,500 feet (8991.6 meters) over the Himalayan Mountains. Others make nonstop journeys that would give us jet lag for a month—Blackpoll Warblers take off from the Massachusetts coast in the autumn and fly out over the Atlantic for 36 hours to a point where they pick up the trade winds of the West Indies, the currents of which carry them to the coast of South America. It is a four-day nonstop trip.

The full, astonishing extent of bird migrations began to become clear in the mid-nineteenth century as the collecting of exotic birds became popular among the wealthy in Europe and America. Snipers were dispatched on expeditions to shoot rare specimens, which were then stuffed and mounted. The feathers of exotic birds became the rage for women's hats, dangerously depleting the numbers of many kinds of large birds, a development that led in turn to the first efforts at protecting birds. The Audubon Society, founded in 1905, led the way, and President Theodore Roosevelt created the first National Bird Sanctuary at Pelican Island in 1907.

During the nineteenth century, although much of the interest in birds had been acquisitive, there had also been developments on the scientific front, chief among them the publication, between 1827 and 1838, of John James Audubon's magnificent *Birds of America*. His paintings depicting native species as he had observed them in their natural habitats was an artistic and scientific achievement of the highest order. Charles Darwin's *Origin of the Species,* published in 1858, was deeply influenced by his studies of birds during his five-year voyage on the *H.M.S. Beagle*. In many ways, his theory of evolution only deepened the mystery of bird migration. If some birds evolved into new species in isolated areas, why should other birds fly so far in order to seek out a winter food source?

Pelicans fly in formation, returning to Pelican Island in the Indian River, Daytona Beach, Florida. Pelican Island is the northernmost nesting ground of these birds in the United States; these pelicans winter in Venezuela. The island was the first wildlife refuge in America, so designated by President Theodore Roosevelt in 1907. Photograph by the author.

At times it seemed that the more naturalists learned about birds, the more confused the situation became. It was not just the incredible distances traveled by some birds that gave scientists pause. What troubled them even more was that the patterns of migration were riddled with inconsistencies from one species to another. For example, most species seem to go out of their way to avoid flying over open water for extended periods. While this seemed perfectly logical, in that land birds have no place to stop for rest and "refueling" over open water, why then do some birds undertake such a difficult journey? How can Blackpoll Warblers spend four days over the ocean without stop? Even more per-plexing, why do Ruby-Throated Hummingbirds—which already have to consume enormous amounts of food to sustain the extremely rapid beating of their wings—make the long journey across the Gulf of Mexico from the southern United States to the Yucatán Peninsula and back? Of all the species, they would seem the most likely candidates to take the long way around, by land, instead of flying 500 miles (804.5 kilometers) across the gulf.

Such baffling questions left many experts doubting that they would ever comprehend the mysteries of bird migration. During the first several decades of the twentieth century, progress was made in establishing the patterns of bird migration, however, as the practice of "banding" or "ringing" the legs of birds in their nesting areas became more common. Helped by small armies of bird-watching enthusiasts around the globe, who would report sightings of banded birds, scientists were able to draw complex maps of bird movements. The *where* and *when* of bird migration was at last being set down in detail, even if the *how* remained elusive.

It became apparent that the majority of bird species did not travel as families. In most cases, the males left the summer nesting area before the females and the fledgling offspring—sometimes months earlier. Male Ruby-Throated Hummingbirds begin the return trip to Mexico as early as the end of July, whereas the females and young birds remain in the United States until as late as October. Three species of swans, on the other hand, including the small Whistling or Tundra Swan and the much larger Trumpeter Swan, both North American birds, migrate as families from their nesting grounds in Alaska and Canada to their winter feeding grounds across the United States. This family togetherness has been attributed to the fact that swans mature more slowly than most birds, and thus the offspring need all the help they can get in finding their way along the migratory route.

This explanation, in itself, raised another, more difficult, question. Why do some birds appear to have an innate sense of migratory routes, while others seem to need far more in the way of parental guidance? Such differences among species suggest an uncomfortable answer: Different species use different kinds of navigational systems to get from one place to another. If that is true—and there is considerable agreement on this point—then a real understanding of bird migration is not a matter of developing a single theory that more or less explains bird migration in general but rather looking at a wide array of different navigational modes.

Since the early 1970s, scientists have indeed suggested just such a variety of guidance systems. Because avian experiments can be very complex, researchers have almost always concentrated on single aspects of the problem. That has meant the development of a number of competing theories.

There is a basic agreement among all researchers that birds, like mammals, are governed by circadian rhythms. The word *circadian* is compounded from two Latin words: *circa,* meaning "around," and *dies,* meaning "day." Birds and humans alike have an internal clock that is attuned to the 24-hour revolution of the Earth. It seems likely that this attunement is even stronger in birds than in humans. If birds are subjected to a sudden shift in the duration of daylight hours in controlled environments, it takes them two or three days to adjust, and their habits and sleeping cycles will be thrown slightly off-kilter during that period. After this adjustment, however, their internal clocks reassert themselves and operate according to a 24-hour cycle regardless of the external stimuli.

The existence of this very accurate internal clock is widely believed to be crucial during bird migration. In the course of the Earth's daily rotation, a single location will swing through 15° of longitude each hour (15° × 24 hours = 360°). Thus, an error in timekeeping will result in veering 18½ miles (29.6 kilometers) off course for every misjudged *minute.* Clearly, circadian rhythms alone cannot possibly explain the pinpoint accuracy of birds in returning to the same nesting area each year after flights of hundreds, even thousands of miles (kilometers).

Another obvious element of navigation is the acuity of avian sight. Have you ever walked out into your garden to be startled by a Blue Jay swooping down out of nowhere to grab a field mouse at your very feet, which you had not even noticed? Such experiences dramatically attest to how much most birds can see from considerable distances. The expression "bird's-eye view" suggests another aspect of the part that superior vision plays in bird navigation. Many experts suspect that even the remarkable ground details revealed by satellite photography fail to match the precision of what many birds can see when in flight.

There are problems, however, in attributing too much import to avian sight. While it is clear that birds use their very acute eyesight in hunting for food and navigating in the nest area, there is scant evidence that landmarks on the ground are of much import during long-distance flights. Experiments in this area are very difficult to control, but the work that has been done has left researchers dubious that visual landmarks play a large role in migration. Given how suddenly the landscape can change due to an earthquake, flood, or forest fire—or human construction—any great dependence on visual cues during long flights would be more likely to create confusion than anything else.

But if landmarks on the ground are not that important, what about signposts in the sky? What part might the position of the Sun, or the stars, play in migration? There is evidence that birds, like humans, must avoid looking directly into the Sun, but it has also been established that pigeons make use of the shadows they cast on the ground as they fly. Back in 1968, a leading authority in birds, Geoffrey Mathews, made a theoretical case for the ability of birds to use a kind of "sun compass." This theory calls for an avian ability to calculate angles and planes that might glaze the eyes of the average high school student, but it is not beyond the range of possibility. Instinctive mathematical gifts of a special kind are evident throughout the natural world, from beehives to beaver dams. Another theory that has received much attention is J. D. Pettigrew's suggestion that the *pecten oculi* within a bird's eye may act like the gnomon of a small sundial, projecting a shadow on the back of the eye, which could be used as a navigational aid.

Even if such theories could be proved, they would still leave us with many unanswered questions. Many birds prefer to fly at night during migration, and others fly day and night nonstop, suggesting that many species must also be able to use the stars to orient themselves. Experiments have been going on since the 1940s with respect to this possibility. In a celebrated example, Stephen Emlen placed newborn Indigo Buntings in cages in a planetarium, and then he manipulated both the placement and the rotation of the projected stars in a series of experiments. It became clear that the "star compass" of the Indigo Bunting was learned rather than innate, a matter of familiarity rather than instinct.

That makes sense because the positions of the stars change over time, and if birds had evolved with a fixed, innate star compass, evolution would have found itself in a constant race with the changing sky, albeit a race taking place over several millennia.

The most successful experiments relating to bird migration have focused on their ability to make use of readings of Earth's magnetic field. The fundamental experiments of this kind were carried out by a team of German scientists in Frankfurt in the late 1970s. Many other research projects have followed up on that work, and taken together, they make it clear that birds do respond to magnetic fields in remarkable ways. The most intriguing development in this area was the discovery of the presence of a tiny magnetic crystal in the head of pigeons, located between the skull and the brain. It has not yet been shown to exist in more than a few species but is at least a possible biological feature that could give birds a kind of "sixth sense."

Theoretically then, there are a number of possible explanations for the age-old questions about bird migration: circadian rhythms, visual acuity, the ability to make use of a sun or a star compass, even a biologically determined sixth sense. The problem is that while a case can be made for all these guidance systems, their importance appears to vary considerably from species to species. Other factors, from the smell of fish to the sound of frogs, also appear to play a role for some species. The great differences among species, in terms of everything from nesting habits to the kinds of food consumed, suggest that quite different mechanisms, or combinations of them, may be at work in a hummingbird and a tundra swan. A species that migrates a few hundred miles (kilometers) may have no need of navigational systems essential to another that covers thousands of miles from the Arctic to the Antarctic and back. Those adorable denizens of the Antarctic, the penguins, also migrate, as much as 300 miles (482.7 kilometers)—but they walk! For them there is no bird's-eye view available, and they have little need for a crystal that reacts to the Earth's magnetic field.

Sixty years ago, researchers despaired of ever understanding how birds were able to complete their extraordinary journeys from locations hundreds or thousands of miles apart with such pinpoint accuracy, but much progress has been made since then. Partial explanations abound, but every book or scientific article on bird migration is full of conditional words and phrases: "It may be . . . but it also might not be." We know more about how birds *might* achieve their epic flights around the world, but there are still far more mysteries than there are explanations. The tiny songbird that reappeared to build its nest in the apple tree outside your window—and we know from banding that it can indeed be exactly the same bird—has been to South America and back since you saw it last.

How can that be?

This is one case where it may be nicer not to know—simply allow yourself to be swept up by awe and wonder.

⚛ To investigate further

Mead, Chris. *Bird Migration.* New York: Knopf, 1985. This wide-ranging account of bird behavior, including migration, is illustrated with more than 600 splendid color photographs of birds in their native habitats.

Elphick, Jonathan, Ed. *The Atlas of Bird Migration.* New York: Random House, 1995. This fine guide to the migratory routes of birds around the globe includes clear maps and much fascinating information. Chris Mead (see previous item) was the chief consultant on this book.

Wade, Nicholas, Ed. *The Science Times Book of Birds.* New York: Lyons Press, 1997. A collection of articles from the weekly "Science Times" section of the *New York Times,* these 60 pieces cover everything from the family life of birds to efforts at bird protection. Five stories on migration are included.

Martin, Brian P. *World Birds.* Enfield, England: Guinness Books, 1987. Although it contains a great deal of scientific information, this book is angled toward trivia enthusiasts, with dozens of entries on the highest-flying, fastest-swimming, longest-billed, and so on, champs of the avian world.

Weidensaul, Scott. *Living on the Wind: Across the Hemisphere with Migratory Birds.* New York: North Point Press/Farrar, Straus and Giroux, 1999. The author traveled some 70,000 miles (112,630 kilometers) over a six-year period in the course of preparing this book, and he has combined scientific accuracy with fine writing to produce what may become a new classic in the field.

Chapter **12**

What Is Red?

Adiver is swimming underwater around a coral reef. In a moment of carelessness, he scrapes his arm against the sharp coral. It isn't a bad cut, but it does bleed. The man looks at his blood. It's green.

It is early June in the northeastern United States. The leaves of the trees are fully developed now. As we look out the window at them, they are as green as can be, many shades of green, bright in the sunlight. See that squirrel there, scampering through the branches? To the squirrels the leaves are not green, but red and yellow.

What is going on here? Blood is red, as we all know. Also, in the northern hemisphere, the leaves of trees are green in June; they won't turn yellow and red until autumn, when colder weather strikes and they begin to die, or "turn," as we prefer to say. Is the color in the leaf, or in our heads? Does the leaf have an intrinsic color that is always there, depending on the season, or do our brains assign it that color based on other kinds of information?

Most of us have encountered the old philosophical conundrum about the tree falling in the forest: Does it make any sound if there is nobody there to hear it fall? We can also ask whether the leaves of the summer tree are really green if nobody is looking at them. Let's make the question about the color of the leaves a little more specific. Are they green in summer if no human being is looking at them? It's already been stated that a squirrel will see them as red or orange. In fact, if the human being look-ing at the leaves happens to be color-blind, the leaves won't be clearly green, either. True *monochromatic vision,* in which every-thing seems to be black, white, and shades of gray, like a 1930s movie, is very rare, but *anomolopia,* the inability to distinguish the difference between red and green, is far more common. In Europe and North America, about 1 male in 12 has some degree of color blindness. This deficiency is genetic and sex-linked, occurring far more frequently in males; only about 1 female in 200 is affected. Even people who are not technically color-blind can have prob-lems, however, because what colors we see also depend on how well we learned the range of colors and their specific names as children. Quite a number of people, women as well as men, have some difficulty in correctly identifying colors because they learned them wrong to begin with. Almost everyone has gotten into an argument at some point with a person who says, "That's blue, not green" or "That's orange, not red." This usually occurs when dealing with colors that are subtle shades, such as a green-ish blue or an orange-tinged red. The argument can often be set-tled by showing a strong blue and a strong green, or a strong orange and a strong red, next to the in-between shade. "Well," the doubter will admit, "I guess it is more green than blue." Even then, there may still be problems, ones that go well beyond the sheer stubbornness some people display in this kind of argument.

The color vision that we take for granted is in fact one of the pinnacles of evolution. Early forms of life on Earth did not even have eyes, only clumps of cells that were sensitive to light. Such "eyespots" are still found among smaller species with soft skins, such as worms. These eyespots serve a dual purpose, assisting in the search for food and warmth while warning of excessive heat that might parch a worm's sensitive skin. Worms are inverte-brates, of course, with no backbone, like more than 95% of the extant animal species. There are only 41,000 vertebrate species, including mammals, birds, reptiles, amphibians, and fish. The dis-tinction between vertebrates and invertebrates was originally based solely on the presence or absence of a backbone, but since the 1960s, another highly significant difference has been deter-mined. All invertebrate eyes evolved from the skin, one reason why starfish, for example, have eyes at the tips of their legs (the points of the stars of these creatures, which are not in fact fish, in spite of the name). All vertebrate eyes, however, are an outgrowth of the animal's brain, a connection that becomes increasingly important as we move up the evolutionary scale.

Detecting light, as even a worm can, is the most basic of eye functions. At a more evolved level, eyes can detect motion, and at the highest level, they can form images, although the ability to do that varies greatly even among different species of mammals. It is crucial to understand, however, that even we humans do not actually see objects. What we see is the light they reflect. Each human eye has more than 180 million receptors that can catch reflected light and begin the process of turning it into images. As the *photons*, the elementary particles of energy by which light is transmitted, stream into our eyes by the millions every second, some are trapped by the clusters of photoreceptors that form the retina (which means "little net" in Latin). The curved transparent cornea of the vertebrate eye behaves as a fixed lens that guides the streaming photons toward the retina behind it.

The retina itself has two separate kinds of receptors, photosensitive cells called rods and cones. The two respond to light in different, and mutually exclusive, ways. Cones are activated by bright light, and they exist in three varieties, one that absorbs light on the blue wavelength, one that absorbs on the green wavelength, and a third that absorbs yellow, making it possible for us to see colors. (The human brain is able to block some of the light on the green and yellow wavelengths, enabling us to perceive red, which has a shorter wavelength that is often invisible to us, as in infrared light.) Rods take over the job of seeing when the light is dim. If you go from bright sunlight into a dark room, or vice versa, you will have difficulty seeing at first because the switchover between the two kinds of receptors does not happen instantaneously. When stimulated by light, cones or rods emit an electrical impulse, which results in a message being sent along the optic nerve for each eye to the visual cortex at the very back of the brain.

Leonardo da Vinci, who was ahead of the game by centuries in a host of fields, was the first to grasp the idea that images were

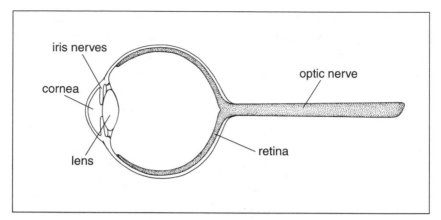

The human eye is a direct outgrowth of the brain. Light strikes the cornea—the clear front window of the eye—which transmits it to the transparent lens, where the light rays are further focused onto the retina, the nerve layer at the back of the eye. The resulting impulses are then sent through the optic nerve to the brain, where visual information is processed in a complex, if seemingly instantaneous, series of steps that are not yet fully understood.

sent from the front of the eye to the retina but actually formed not in the eye itself but in the brain or imagination (both being implied in the Latin term *sensus communis*). As in so many other things, he was right. What actually happens in the brain's visual cortex, however, is extremely complicated and not well understood. The visual cortex is known to be divided into well over 20 areas, but only the half dozen that handle the signal first have begun to be thoroughly investigated. The results are complex, with the incoming signals being analyzed, compared, and sent back and forth for corrections before being registered as an image that the individual finally "sees." All of this happens in a way that seems to us instantaneous, although a person who is drunk or injured may find the process considerably disrupted.

In other vertebrates, not to mention invertebrates, evolution has produced biological mechanisms that can be very different from our own. It is hardly surprising that the vision of other primates, particularly the African great apes, most closely resembles our own. After comparisons with fellow primates, however, our understanding of vision becomes confusing, even among mammals. The giraffe has been tested for color vision and seems to be able to recognize reds and violets, but the giraffe has trouble distinguishing among green, orange, and yellow. Red-green color blindness is very common in mammals—that squirrel in the tree, for instance. The color vision of dogs is limited, though it is compensated for by very keen senses of smell and hearing. Cats have a greater degree of color vision than dogs, but the colors are pale. You may have noticed how the pupils of cats narrow greatly in bright light—that is because they have so few cones in proportion to rods that they must block out as much light as possible to keep the rods from shutting down altogether, and this process also washes out the color. The differences in vision between one mammal and another may be fairly slight or very marked, but no two species see in the same way. Evolution has decreed that each lives in its own world. That world literally *looks* different to each species.

The differences are even greater between creatures that have a single-lens eye like our own, and those that have compound eyes. Single-lens eyes are associated with the need to see clearly into the distance (to detect the presence of both prey and predators), and all vertebrates have them. Compound eyes are designed for seeing well at close range, and most invertebrates have them, although the king crab and a few other species have both single-lens and compound eyes, and worms, as noted, have only eye spots. The tiny lenses of compound eyes can range from 10 in a ladybug to thousands in a bee—and that bee can detect details of a flower that we would need a microscope to see. Experiments have shown that most flying insects see in color, and some butterflies in particular see a range of colors beyond any other living creature. It is believed that the spectacular color vision of some flying insects is directly related to the fact that flowering plants and insects evolved contemporaneously—a point we come back to later in this chapter.

Underwater vision presents special evolutionary problems. To start with, as the photons that carry light pass through water, they are scattered as they hit molecules of water, creating glare. Water also acts as a filter and blocks out more and more wavelengths of light at increasing depths. Ultraviolet and infrared wavelengths—which humans cannot see anyway—are blocked below 50 feet (15 meters), and only blue-green wavelengths penetrate farther down, which is why the injured diver will see his blood as green. A fish that appears red when brought to the surface will appear blue-gray (to us) in even deeper water. Whereas the most primitive fish, such as sharks and rays, which have been around for millions of years and are held together by cartilage rather than bone, do not have color vision, the bony fish that evolved later do have it. You may think it strange that color vision would develop when all fish in deep water appear blue-gray, but keep in mind that that is the color we see. Fish have very fine-tuned vision in the blue-green spectrum and can see color differences among species that are invisible to us. Once again, these animals live in a different world that has color variations beyond what we are able to detect.

Finally, there are the birds, so often brightly colored and with color vision that is superior to ours. There are some surprises here, too, though. The glittering iridescence that we so admire in hummingbirds is not color at all. The feathers are gray, but the incessant rapid beating of the hummingbird's wings keeps those feathers in motion, and as light travels through the nearly transparent layers, it creates an entirely illusory special effect. Because of the way our eyes process color, we are seeing something that is not really there. In addition, the extraordinary eyesight of birds, in terms not only of its color range but also the distances it can see (hawks can see eight times farther than humans), comes at a price. The complexity of a bird's eye takes up a great deal of space, filling almost the entire skull of large-eyed owls, which means that there is not much room left within the skull for a brain. ("Wise as an owl" is a misapprehension based on the gravity of their appearance and their habit of sitting very still for long periods.) As we saw in chapter 11, birds are able to migrate thousands of miles even with their small brains, which are adapted to such tasks regardless of their size. Nevertheless, the eye of the bird has evolved at considerable cost to its brain.

Yet human beings also have remarkable color vision, second only to birds and some insects in the range of its spectrum, and our brains are very large. How did this unusual combination happen? Because we are far from understanding how the visual cortex of the brain itself works, many of the answers to that question clearly lie in the future—if indeed they can ever be fully pinned down. Nonetheless, there is considerable agreement among both vision specialists and evolutionary biologists that the human brain and human eyes developed in tandem. More precisely, the eyes of primates periodically evolved to a new level of acuity and color perception, and this improvement goaded the brain to further enlargement in order to successfully process the new level of information provided by the streams of protons the eyes were receiving. At times, the growth of the brain might in its turn propel further developments in vision. This process has a parallel, it would seem, in the extraordinary color vision of butterflies, which evolved simultaneously with flowering plants. That development was an interaction between two utterly different forms of life, however. In the case of primates, a simultaneous interaction occurred within a single organism. This took place over millions of years in tiny increments, of course, and there is no doubt that the final surge of brain development that created human beings was affected by many other factors, particularly the changeover to walking erect on two legs. By the time bipedalism developed, it is believed, the eyes had also developed sufficiently to make such further steps possible.

The question still remains: What is red? Is blood red or a leaf green in essence or only in our minds because of the way our brains process light? If blood is green when spilled from a cut underwater, and the summer leaves are red and yellow to the eyes of squirrels, can we say with any finality that these colors are intrinsic to blood and to leaves? Instead, is all of the visible world to some extent an illusion, like the iridescence of hummingbirds, a matter of how a given brain has evolved to interpret the photons that stream from our sun and every star in the galaxy? We do not see objects, after all, but waves of light. In our world, to us, the grass is green, but if intelligent beings from another place, a planet in a system with two suns, for example, beings whose eyes had developed to make sense of the very different kind of light produced in such circumstances—if such beings landed on Earth, would they say that the grass is green? It might be deep blue or even red to them.

And what is red?

✳ To investigate further

Birren, Faber. *Color and Human Response.* New York: John Wiley & Sons, 1997. This book is aimed at general readers and lays out the basics well. It is also primarily a book about psychological responses to color, with chapters on such subjects as "To Heal the Body" and "To Calm the Mind."

Lauber, Patricia, with photographs by Jerome Wexler and Leonard Lessin. *What Do You See and How Do You See It?* New York: Crown, 1994. Parents might like to sneak a look at this book for "young readers." It manages to get across a lot of information clearly and without too much oversimplification, and it describes optical experiments that kids can try for themselves.

Sinclair, Sandra. *Extraordinary Eyes.* New York: Dial, 1992. This is another excellent book for young readers.

Note: Vision is such a technical subject that the best recourse for the general reader may be articles in magazines, and not just science magazines. One of the most interesting articles on vision in recent years appeared in *The Economist,* April 3–9, 1999. Titled "The Biology of Art," and written by the editors, it discusses how various schools of art, from impressionism to cubism to the mobiles of Alexander Calder, draw on and play with specific biological aspects of human vision. At this writing, it was available on the Internet at britannica.com.

Chapter 13

How Did Mayan Astronomers Know So Much?

W hen the Spanish *conquistador* Hernán Cortés first marched on the Aztec capital of Tenochtitlán (now Mexico City) in 1518, neither he nor his men seemed to have any appreciation of the sophistication of the culture to which they would soon lay waste. The Spanish noted with disdain that despite the Aztecs' grand architecture and the gold that adorned their leaders, these New World "Indians" didn't even have wheeled vehicles. Nor were later explorers and governors much impressed with the remnants of the older Mayan culture that survived in the jungles of the Yucatán Peninsula. Some Roman Catholic priests did understand something of what they were seeing, and a few were sympathetic, but the most important Spanish administrator in the Yucatán during this period was a Franciscan friar named Diego de Landa, who had been schooled by the Inquisition. He was determined to stamp out the Mayan religion, not only destroying representations of the Mayan gods, but also burning a large depository of ancient hieroglyphic manuscripts. Ironically, de Landa was as fascinated by the Mayan culture as he was determined to destroy it, and he made copious notes, which he later used to write a treatise on the Maya in the 1560s. A copy of that treatise, with some sections missing, was rediscovered in Madrid 300 years later, in 1863. It served as the key to deciphering the three undisputed ancient codices of the Maya that had survived despite de Landa's slash-and-burn tactics.

The most important of these manuscripts is the Dresden Codex discovered in Vienna in 1739. It is assumed that it had originally been brought back from the Yucatán as a kind of souvenir. None of the fully authenticated surviving codices deals with Mayan history—a loss that haunts scholars to this day—but the Dresden Codex reveals the astonishing degree of astronomical knowledge among the Maya, while the Codex Tro-Cortesianus focuses on ritual and prophecy, and the Codex Peresianus details ceremonies connected to the extremely complex Mayan calendar.

It was not until the end of the nineteenth century that the meanings of these codices began to be unraveled by various scholars around the world, particularly Léon de Rosnay of France, Cyrus Thomas of the United States, German philologist Ernst Föstermann, and ultimately, in 1897, a California newspaper editor named Joseph T. Goodman, who borrowed liberally from the German's work without giving credit. Goodman made up for this lapse in 1905 by working out the correlations between the Mayan and Christian calendars, an achievement of importance to Mayan studies ever since.

It became apparent that the planet Venus was of vast significance to the Maya, and essential to the structure of their complex calendar. The primacy of Venus is not in itself surprising. As the brightest object in the night sky, aside from the Moon, the "evening star" had been the focus of religious observances in numerous ancient societies, starting with the Sumerians, as early as 3000 B.C. Although several cultures had used Venus as a kind of lodestar, none had approached the accuracy of observations of this planet achieved by the Maya. During what is called the Classic period of Mayan civilization (a designation intended to evoke associations with classical Greece), from A.D. 300 to 900, the Maya had developed methods for keeping track of the revolutions of Venus that would not be matched by European astronomers until telescopes became prevalent in the eighteenth century. Indeed,

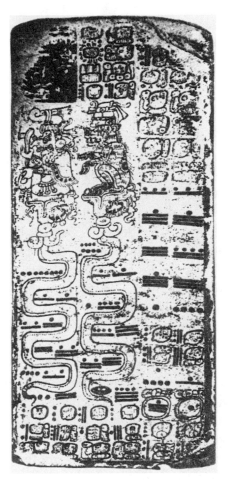

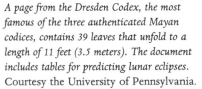

A page from the Dresden Codex, the most famous of the three authenticated Mayan codices, contains 39 leaves that unfold to a length of 11 feet (3.5 meters). The document includes tables for predicting lunar eclipses. Courtesy the University of Pennsylvania.

their figure of 584 days for a transit of Venus through the sky is almost exactly the same as that arrived at by astronomers using modern instruments: 583.92 days. How could they have achieved such accuracy centuries before the supposedly far more advanced scientists of Europe were able to do the same—and why did they do it?

Surprisingly, the second is the more easily answered question. The Mayan concern about the passage of time, and its meanings, verged on the obsessional. As Anna Benson Gyles and Chloe Sayer write in their 1980 book *Of Gods and Men: The Heritage of Ancient Mexico,* "Past and future receded in endless vistas of hundreds and thousands of years, as the ancient Maya tried to measure the passage of time and to solve its mysteries. By observing the length of the lunar cycles, the equinoxes and solstices, the revolutions of the planet Venus and the passage of the seasons, the Maya were able to evolve their own highly elaborate and precise systems for recording time."

All of the Mayan astronomical observations—Mars and Jupiter were also included in these to a lesser extent—were put to use in keeping three separate calendars of the year, a longer-term calendar that united the others, and a "time-line" calendar that reached far back into the past. Of the three basic calendars, the *tzolkin* calendar consisted of a 260-day sacred year. It is believed that it may have been inherited from the earlier Toltec culture. The second calendar, called the *tun*, was divided into 18 months of 20 days each, adding up to 360 days. A third yearly calendar, the *haab*, conformed to the 365 days we know, but the extra 5 days, which were considered unlucky, composed a special month in themselves. While this may sound confusing, and indeed like bad math on the surface, it was all pulled together in what was known as the Calendar Round.

The Calendar Round involved keeping track of 18,980 days, all of which had individual ritual significance. If their 260-day year is multiplied by 73, it will amount to 18,980 days, which is exactly equal to 52 actual 365-day years. Thus, a new cycle began again every 52 years, and there is evidence that Mayans from cities as much as 300 miles (480 kilometers) apart from one another traveled through the Yucatán jungle to meet in a central place at the end of such cycles to make certain that all their calendars were properly synchronized.

Surprisingly, that feat of number-crunching was nothing in comparison to what is known as the "Long Count," which is regarded as the most accurate calendar devised by any ancient culture. Charles Gallenkamp lays out the entire sequence in his 1985 book *Maya*. It begins with the *kin,* or single-day unit:

20 kins	=	1 uinal (20 days)
18 uinals	=	1 tun (360 days)
20 tuns	=	1 katun (7,200 days)
20 katuns	=	1 baktun (144,000 days)
20 baktuns	=	1 pictun (2,880,000 days)

The count continues for three more levels, reaching the astounding total of 1 *alutun,* or 23,040,000,000 days. As is obvious from these numbers, the Maya had a symbol for zero, a somewhat abstract shell shape. Only three cultures in the history of humankind managed to invent zero: the Babylonians, the Hindus, and the Maya. The Babylonian zero lapsed into disuse with the fall of that civilization, and it was not until the ninth century A.D. that the Hindus invented it again, making modern mathematics possible. Even so, the zero was not introduced into Europe until the Middle Ages, where it took hold as the clumsy Roman numeral system at last died out. Here, too, the Maya were a thousand years ahead of the "civilized" world from which their final destroyers would eventually set sail.

The Mayan calendar had been calculated backward to a starting date regarded by scholars to be 3113 B.C. This was apparently a Creation date. The Long Count was exact to a single day out of a 6,000-year period, a fact established through the relatively recent use of computers, of course. Even more astounding than the sheer mathematical genius of this ancient people is that each day had a particular significance. They believed that each succeeding day was carried on the back of a different god, in an endlessly repeated cycle. Many of these gods were good, but some were bad, and on days when a bad god was known to be carrying the day (including the unlucky five-day month each year), unfortunate things were likely to happen. What was more, certain periods of certain years during the Calendar Round were particularly dangerous, as was also true of periods in the Long Count.

The astronomers and mathematicians among the Maya were a priesthood, and a powerful one. Their keeping of the cosmic books informed the populace, peasant and ruler alike, of dangerous days or conjunctions of calendrical time, as well as the best times for planting and other activities of daily life. It is clear that there were an enormous number of rituals in Mayan life, although few of them have been clearly understood. Although these rituals were performed in order to appease or please the gods that carried the days, it has become increasingly clear that the Maya were a deeply fatalistic people. They believed that time was cyclical, as their 52-year Calendar Round makes obvious, but there were also more complex conjunctions within the Long Count, which were seen as inevitably recurring.

The fatalism of the Maya is one of many possible explanations offered for the collapse of the civilization around the year 900. Scholars have staked out a number of different positions on why things went so wrong so suddenly. One theory holds that as the population surrounding the major Mayan cities increased, it became impossible to feed them all. Other theories posit disaster scenarios involving earthquakes or hurricanes, and disease scenarios that apply to either humans or the plants, particularly maize, that the Maya grew. There is evidence that the Mayan city-states were rivals and often engaged in wars with one another, which could have eventually depleted the resources of all sides. There have been suggestions that a peasant uprising against the rulers and priests could have triggered the collapse of Mayan civilization. No one really knows. One of the most interesting theories is that whatever the surface cause—earthquake, disease, uprising— it came at a time when the calendrical conjunctions promised disaster. There is good evidence that a time of terror had been predicted—and because it was an inevitable part of the great cycle of time, the rulers and priests regarded resistance as futile. In this view, the end of the great Classic period was an inevitable by-product of the very mathematical and astronomical sophistication that fed their obsession with time in the first place.

Although we can understand how their religion and view of the world could have spurred the Maya to feats of mathematics and astronomy far in advance of what was happening in Europe in the period from A.D. 300 to 900, we know little about how they did it. Although the placement of some Mayan temples and other buildings suggest relationships that could be astronomical in nature, particularly with respect to observing the solstices, there is nothing so clear-cut as Stonehenge in England. There are slits in walls, which could have been used as astronomical devices. Yet they often seem slightly off-kilter, which doesn't make sense in a culture with calendars of such astonishing accuracy. Given the complexity and beauty of Mayan carvings, it would seem that they had the capacity to carve smaller devices that could be used in observing Venus, for example, but there are none to be found.

The first Europeans to explore Central America were often impressed with the great architecture they saw, but they were very perplexed by the absence of wheeled vehicles. That lack was just one of many things that contributed to the view that the Aztecs and the scattered Mayan survivors in the Yucatán were "savages." We know now that the Maya understood the wheel. Archeologists have found miniature wheeled vehicles, replete with axles, that were clearly children's toys. It may have been that in the damp, hot jungles of the Yucatán, wooden wheels sank into the ground, or rotted so quickly that they were regarded as useless. That impermanence of wood could also be a clue to the astronomical devices of the Maya. Perhaps, some experts think, they were made of wood. They could even have been little more than crossed sticks marked with numbers. If so, they would certainly have rotted away by the time the Spanish arrived six centuries after the abandonment of the great pyramid cities had brought an end to the astronomer-priests who knew so much about the planet Venus—but that is just speculation.

Chapter 8 noted that the earliest forms of language and mathematics, in Sumeria and Egypt, appear to have been clearly connected to the collecting of taxes. The need to make an accounting was the spur to writing some 3,000 years before the birth of Christ. It is to that period that the Long Count of the Maya takes us back, too. There is little question, in the minds of most scholars, that some of the knowledge on which the Maya built such an extraordinary mathematical superstructure goes back even further in the pre-Columbian world of Central America. It might even be speculated that the base date of the Long Count, 3113 b.c., according to scholarly conversion to the Christian calendar, was not seen as a creation myth in the usual sense, but rather might refer to the beginning of mathematics. Yet, to judge by what the Maya did with the mathematics and the astronomy they practiced, numbers were seen as necessary for a quite different reason than in Sumeria and Egypt. Mathematics was not for keeping an account of worldly goods, but rather for keeping track of which of the gods was carrying time on his or her back each day. This need provided the inspiration for mathematics and astronomy in this very different world across a then-unknown sea.

We have also seen, in chapter 1, that some physicists feel that contemporary efforts to understand the beginning of our universe are tending toward a kind of theology. That higher mathematics, of an intricacy that few people can understand, should be put to the use of divining the origins of everything, and that that effort should have religious overtones in some ways, seems less odd in light of the world of the Maya. Numbers and gods, for the Maya, were inextricable. For that very reason, of course, it can be asked whether their astronomical observations or the immensely sophisticated calendars they devised can even be considered "science" as we think of it today. Because we do not know how their astronomical observations were made, some may feel that the Maya merely stumbled onto a measurement of the transit of Venus, the import of which they did not actually understand. On the other hand, they did know things that Europeans who regarded themselves as scientists did not, and they established the movement of Venus through the sky with an accuracy borne out by computers in our own time.

In the next two chapters, we will be meeting Sir Isaac Newton, whose theories and experiments made him the father of the modern scientific method. As we encounter him using wooden boards, a prism, and the light from a window to reveal the secrets of light (chapter 15), it is worth pondering the relationship between what he was doing and what the Mayan astronomer-priests were up to with their possible crossed sticks held up against the night sky. There are certainly differences, but perhaps likenesses, too.

❋ To investigate further

Gallenkamp, Charles. *Maya*. New York: Viking Penguin, 1985. This is the third revised and updated edition of a book originally published in 1959. That is considerable longevity for a book of this kind, and it remains, even today, an excellent introduction to the subject. Gallenkamp served as coordinator for the exhibition of Mayan treasures that toured American museums from 1985 to 1987. It should be noted that there was an explosion of knowledge about the Maya from the 1950s to the early 1980s. Although scholarly articles continue to appear on a regular basis, almost all the solid books for the general reader date from 1985 or earlier.

Gyles, Anna Benson, and Chloe Sayer. *Of Gods and Men: The Heritage of Ancient Mexico.* New York: Harper & Row, 1980. Another older book that is still worth reading, this one has fine black-and-white illustrations on nearly every page. It focuses to some degree on the relationship between modern Mexico and its heritage, and the authors have done their research well.

Krupp, E. C. *Echoes of the Ancient Skies: The Astronomy of Lost Civilizations.* New York: Harper & Row, 1983. While this book doesn't deal with the Maya in any depth, its overall look at the beginnings of astronomy in early cultures around the world is fascinating.

Chapter 14

What Is Gravity?

Every schoolchild knows the story about the guy in funny-looking clothes sitting under a tree and being hit on the head by an apple: "Wow," says Isaac Newton, "there must be something called gravity." Of course, it was a bit more complicated than that. Galileo had recognized that an apple and a melon, for example—two objects of different sizes and weights—would hit the ground at the same time when dropped from the same height. He spent years working on a law of falling bodies, which he published in his *Discourses* in 1638, four years before Newton was born.

But Newton was to take things much further. He graduated from Cambridge University in 1665 at age 23, and because the cities of England were centers of bubonic plague at that time, he returned to his country home in Lincolnshire. There, over the next two years, he made a spate of breakthroughs, the importance of which would not be equaled until Einstein's great burst of creativity in 1905. Newton's discoveries included differential and integral calculus, the unpacking of white light into its constituent colors (he used a prism to demonstrate the idea), and, above all, the three standard laws of motion and the universal law of gravitation.

It would be 21 years before he published the laws of motion and gravitation, however. He had previously published his findings on calculus, only to find that German mathematician Gottfried Leibniz was staking a claim to this discovery. Newton was convinced that Leibniz had stolen the idea from him, although in fact the German had arrived at the same conclusions quite independently, just after Newton did, something that often happens in science—"an idea whose time has come." Because of this experience, a paranoid Newton sat on his gravitational laws for two decades. His friend Edmond Halley, the Astronomer Royal, finally persuaded Newton to publish, suggesting that he would be genuinely scooped by someone else if he didn't. Halley worked on the manuscript of the *Principia Mathematica* with Newton and paid for its 1687 publication, although he was not a wealthy man. His generosity was repaid with his own immortality, however. Using Newton's universal law of gravitation, Halley was able to work out the elliptical orbit of the great comet that bears his name, and therefore to predict its 76-year cycle of return.

Gravity, as Newton defined it, is the force of attraction that arises between objects by virtue of their masses. The degree of attraction between two large objects is greater than that for two small objects. Also, if two objects are close together, then the attraction between them will be greater than that if they are farther apart. Another way of saying this is that the gravitational force between two bodies is proportional to the product of their masses, but inversely proportional to the distance between them. A ball thrown into the air will return to the ground because the mass of the Earth is so much greater than the mass of the ball. If the ball is propelled to a great height, like a pop foul hit with a bat, it will take longer to come down because the distance between it and the Earth is greater. It is important, however, not to confuse mass with weight. Astronauts bounding along the surface of the Moon still have the same mass that they do on Earth, but their weight is less because the gravitational pull of the Moon is only one sixth that of the Earth. The gravitational relationships have changed: Astronaut/Earth and astronaut/Moon are two different equations governed by the same law of gravity.

It is very difficult for us to fully grasp the initial impact of Newton's ideas because they have permeated science ever since. To make a modest modern comparison, it is like trying to explain

Although the visionary Romantic British artist, poet, and philosopher William Blake was intellectually at odds with the mechanistic universe of Isaac Newton, his several allegorical representations of Newton are among his most powerful works, perhaps because Blake understood the solitary nature of genius even as he reacted against the causalities of Newton's mathematics. Courtesy the Lutheran Church in America, Glen Foerd at Torresdale, Philadelphia.

to a 20-year-old today how revolutionary Jean-Luc Godard's use of the jump cut was in his 1960 film *Breathless*—the technique is now commonplace in moviemaking, although when it was first used, it did indeed take the breath away. Newton's laws did the same, causing the most brilliant minds of the age to gasp at their daring, as well as their ultimate simplicity. He was able to make the connection between the falling apple and the motion of the Moon around the Earth, one falling to the ground while the other remains suspended. Motion, of the right kind in the right direction, can balance, or even overcome, the force of gravity. Both the fact that the Moon stays where it is instead of crashing into the Earth, and the sight of *Apollo 11* escaping the gravity of the Earth to travel to the Moon are accounted for in Newton's laws.

The world Newton revealed to humanity was mechanistic and deterministic. If you know the initial position and the velocity of an object—be it baseball or rocket—you can determine precisely where it will end up. If the baseball drops into center field, just short of a home run, or a rocket fails to reach orbit, it is because sufficient velocity to counteract the pull of gravity has not been achieved. Newton ushered in the Enlightenment, or Age of Reason, as the eighteenth century has come to be called. Human ingenuity had revealed the mechanics of the universe itself—and the place of God in that universe would never be quite the same. Although Pope John Paul II recently "apologized" for the Church's action in forcing Galileo to recant nearly 360 years ago, the Church at the time correctly understood the danger Galileo represented—and he had only gone part of the distance toward determining the laws of the universe.

Newton changed everything, for science and for the way societies were constructed. The American and French Revolutions of the late eighteenth century were an inevitable outgrowth of Newton's explanations of the physical world. Those who understood the motions of the stars had no need of kings to tell them what to do or think. Newton's universal laws had such an enormous impact that by the end of the nineteenth century, many scientists were suggesting that there was no more to be discovered. Electricity, the telephone, photography, the combustion engine—what could be left? There was that nutty idea about building flying machines, of course—but most people were certain that would never happen, even though the laws that govern flight were implicit in Newton's work. As the twentieth century dawned, flight did come, at Kitty Hawk, North Carolina, in 1903. Newton's laws had triumphed again—sufficient velocity had been attained to escape the pull of Earth's gravity. Then, only two years later, another revolution began.

Albert Einstein was a very obscure man in 1905, working as a clerk in a patent office. He published four scientific papers that year, which would change science as much as Newton's laws had in 1687. Only 10 years earlier, when Einstein was 16, his Greek teacher at the Luitpold Gymnasium in Munich had told him, "You will never amount to anything." The boy was thinking about other things than his Greek lessons, as is often the case with the greatest minds. It is doubtful that the Greek teacher would have changed his mind when Einstein's four papers were published in 1905: Very few people read them, and even fewer of those who did were able to understand them—with one notable exception. Max Planck, whose 1900 paper on the quantum theory was further developed by Einstein, read the younger man's 1905 articles and concluded that the Newtonian universe was "dead." Of course, Newton's laws would continue to apply to everyday reality, but Einstein had opened the way to an entirely new kind of universe, one that physicists are still trying to reconcile with Newton's.

First, we return to Newton. There was a problem with Newton's theory of gravity that even he recognized. How did the force of gravity "travel" through empty space? "It is inconceivable," Newton wrote, "that inanimate brute matter, should, without the mediation of something else, which is not material, operate upon and affect other matter without mutual contact. That Gravity should be innate, inherent and essential to matter so that one body may act upon another at a distance thro' a vacuum without the mediation of anything else, by and through which their action and force may be conveyed, from one place to another, is to me so great an absurdity that I believe no Man who has in philosophical matters a competent faculty of thinking can ever fall into it. Gravity must be caused by an agent acting constantly according to certain laws; but whether this agent be material or immaterial, I have left to the consideration of my readers." In short, although gravity clearly existed, what was carrying it?

His readers—or at least the scientists among them—basically decided that the answer lay in an immaterial agent: Space, it was supposed, was suffused with an invisible and frictionless medium that would propel gravity (and light) forward as the ocean does waves. This was called the ether, and it was one of those incorrect ideas, like birds hibernating instead of migrating, that lasted quite a long time because there was no better explanation available. In 1887, however, American physicists Albert Michelson and Edward Morley conducted experiments showing that there was no ether. So it was back to square one: How did gravity retain its force in empty space?

Einstein began to hint at an answer in his special theory of relativity in 1905, and he expanded on it in 1907 when he published his famous equation $E = mc^2$, which indicates that mass and energy are equivalent and interchangeable. The exchange rate—unlike that between the currencies of different countries—never changes. E is energy, and the force of the energy can change; m is mass and that, too, can change, but the conversion rate is always *c squared,* or the speed of light multiplied by itself. Because the exchange rate is so high, an enormous amount of energy can be stored in a very small mass—as the explosive power of an atom bomb makes clear. One implication of this famous equation is that it can take only a small amount of energy, relatively speaking, to create enough velocity to overcome the gravitational force—which is why *Apollo 11* was able to carry people to the Moon. Whereas it took a multistage rocket to lift off from the Kennedy Space Center and escape the Earth's gravity, it took only a modest rocket thrust to lift the landing module back off the Moon again.

The full problem of gravity was not dealt with until the general theory of relativity in 1915. This new theory of gravitation dispensed with any need for an ether. Indeed, Einstein got rid of the Newtonian forces altogether. In Newton's universe, space was static; in Einstein's it was dynamic. According to general relativity, space itself was elastic, and it could be curved, stretched, or seriously deformed by the mass of an object. Our sun would curve the light passing near it because its gravitational field would distort the space in that region. Larger stars would create an even greater distortion, and black holes, it would eventually be recognized, would bend space in almost unimaginable ways. Matter, Einstein showed, warps space.

There was great beauty in Einstein's mathematics—a quality that means a great deal to physicists—but could his idea be tested against an observable event? The answer came three years later when British astronomer Arthur Eddington traveled to Príncipe Island, off the coast of equatorial Africa, to take measurements of the sky during a solar eclipse on May 29, 1919. If Einstein was right, then during the brief period of darkness when the eclipse was complete, there ought to be a distortion in the apparent position of the stars. The distortion not only was there but also conformed almost exactly to the degree that had been predicted by the general theory of relativity. When asked how he would have reacted if the results had been otherwise, Einstein replied, "I would have had to pity our dear Lord. The theory is correct." Although he would come to be regarded by the public as a great charmer, Einstein was not precisely a modest fellow.

Einstein's gravitational theory doesn't invalidate Newton's, however. Newton's "forces" still work in terms of the scale of the solar system—and certainly in terms of home runs, forward passes, and hammers dropped on toes. Newton's theory runs into trouble at larger scales, however, and there Einstein takes over. Newton's theory cannot account for black holes, for example, whose gravitational pull is so great that even light cannot escape them; Einstein's theory neatly takes care of that bizarre situation, because it is the warping of space by the incredibly dense black hole that traps the light.

Newtonian gravity has been downgraded in another way. When his theory was first propounded, gravity seemed to be the most potent force in the universe, one that kept planets and stars fixed in their courses. Yet although it is the glue of the universe, it is in fact the weakest of what are called the four forces. Imagine a baseball game in a major-league stadium in a city for which the electricity is provided by a nuclear-power plant. A ball is hit high in the air toward center field, but the *force of gravity* causes it to fall to Earth just short of a home run. Still, it is enough to bring in the runners already on second and third base, and the scoreboard flashes new numbers. The lights in the scoreboard are made possible by the *electromagnetic force*. The power plant generating the electricity is making use of the *"weak" nuclear force* that governs the disintegration of atoms and thus the radioactivity of the atomic fuel. Finally, the seats the spectators occupy, the hot dogs they eat, the bats and balls on the field, and indeed the spectators themselves are made up of atomic nuclei representing the *"strong" nuclear force*.

On the level of elementary particles the force of gravity counts for almost nothing. The electron and proton that make up a single hydrogen atom are held together not by gravity but by the vastly greater power of the electromagnetic force. How much greater? Ten to the fortieth power (1 followed by 40 zeros) greater. As French physicist and writer Trinh Xuan Thuan puts it, "Suppress the electrical force, and the hydrogen atom, left under the sole influence of the gravitational force, will swell up to fill the whole universe. The force of gravity is so weak that it cannot hold the electron closer to the proton than a distance of a few tens of billion light years."

Only when enormous numbers of atoms are put together do they attain a mass capable of exerting a gravitational force. Mount Everest does not create enough gravitational force to pull a single human being toward it in the physical sense. You can stand at the base and wave good-bye to those brave or foolhardy enough to try to climb it, people who are being drawn to it because of psychological forces within themselves. Those climbing the mountain will be fighting the much greater gravity of the entire Earth, and if they slip, the Earth's gravity will certainly pull them down. Gravity's effects can kill people, but in the larger picture, it is almost negligible. It takes the entire mass of the Earth to keep a piece of paper resting on the top of a desk. Despite being the weakest of the four forces, however, gravity has ironically managed to create the biggest problem in contemporary physics.

Quantum theory, on which the Big Bang theory of the origin of the universe is based, has managed to predict fundamental interactions among three of the four forces: the weak and strong nuclear forces and the electromagnetic force. That still leaves gravity—both Newton's and Einstein's versions—off to one side. Unless gravity can be integrated with the three other fundamental forces, there can be no "theory of everything," or grand unified theory, the holy grail of modern physics. Even to fit the electromagnetic force into quantum physics took many years, largely because it was necessary to develop what are called "renormalization" calculations that cancel out the infinities that have been the bane of modern physics. The late physicist Richard Feynman, a witty man and best-selling author, when asked why he had won the Nobel Prize, once replied, "For sweeping the infinities under the rug."

But renormalization didn't work very well for the force of gravity, which, as David Lindley points out in his 1993 book *The End of Physics*, presents greater complications than electromagnetism. "When two bodies are pulled apart against the gravitational attraction, energy must be expended, and if they come together energy is released; but energy, as Einstein so famously proved, is equivalent to mass, and mass is subject to gravity. Gravity, if you like, gravitates." The relationship between mass and energy, in other words, is an interactive one that has a circular dynamic. This means that the infinities made to disappear by renormalization with respect to electromagnetism are more prevalent and more difficult to "sweep under the rug" with respect to gravity.

The problem ultimately comes back to the same mystery that Newton, in his wisdom, decided to leave to his readers: What is the agent that conveys the force of gravity through the vacuum of space? Many physicists are quite sure where the answer ought to lie: in the existence of a hypothetical subatomic particle that has been given the name *graviton,* a quantum particle equivalent to the photon, which carries light. Both the photon, the existence of which has long since been confirmed, and the supposed graviton belong to the group of messenger particles called "bosons." The graviton ought to exist, and indeed if it doesn't exist, a great deal of rethinking of quantum mechanics will be required.

The search for the graviton is on in a big way. All violent events in the universe, any supernova explosion or collision of galaxies, produce gravitational waves that eventually reach the Earth. Two vast new gravitational observatories, with arms 2.5 miles (3.5 kilometers) long, located in Louisiana and Washington state, have been built to detect such cosmic gravitational waves and to capture them for research purposes. The Laser Interferometer Gravitational Observatories (LIGO for short), it is hoped, will at the very least point the search for the elusive graviton in the right direction. For now, however, the agent that conveys gravitational force remains almost as much of a mystery as it did to Isaac Newton.

✷ To investigate further

Greene, Brian. *The Elegant Universe*. New York: Norton, 1999. This is a well-written and (when the material allows) very clear book about the path that led to the very latest ideas in physics. Greene, a professor at Columbia University, is himself an advocate of "string theory." That material is intrinsically difficult, but he can be very sharp about historical turning points in physics.

Lindley, David. *The End of Physics: The Myth of a Unified Theory*. New York: Basic Books, 1993. This controversial book raises questions about the kind of abstract theoretical directions that Brian Greene (see previous item) promotes. Like many books that challenge cutting-edge physics, it is easier to understand than books by supporters. This could be a matter of oversimplification, or a matter of seeing through a certain amount of double-talk. Greene's and Lindley's books are worth taking a look at in tandem, exactly because they are in opposition.

Suplee, Curt. *Physics in the Twentieth Century*. New York: Abrams, 1999. Published in association with the American Physical Society and the American Institute of Physics, this book has an almost equal balance between text and pictures. The captions to the pictures, which are often spectacular but resolutely scientific, are often as informative as the text. For the general reader this may be the most approachable book on twentieth-century physics yet published—a bang-up job.

Ferris, Timothy. *Coming of Age in the Milky Way*. New York: Morrow, 1988. Winner of the American Institute of Physics Prize, this remains, even a dozen years later, the best introduction there is to the field of cosmology in general.

Thuan, Trinh Xuan. *The Secret Melody*. New York: Oxford University Press, 1995. Thuan is as much poet as scientist and therefore always a delight to read. His elucidation of the four fundamental forces, in terms of a storm in a small town, was the inspiration for the baseball analogy used in this chapter.

Bodanis, David. $E = mc^2$: *A Biography of the World's Most Famous Equation*. New York: Walker, 2000. Among the books listed here, this is the most recent, and perhaps the best, attempt to explain Einstein's famous equation in terms that can be easily grasped by the general reader. Bodanis lays out the earlier work by such nineteenth-century scientists as James Clerk Maxwell, which Einstein built on, and he includes entertaining historical and biographical material.

Chapter **15**

What Is Light?

"Let there be light." (*The First Book of Moses,* called "Genesis")

"Nature, and Nature's Laws lay hid in Night: / God said, Let Newton be! and All was light." (Alexander Pope—intended epitaph for Isaac Newton)

Creation myths from the beginning of human history have tried to account for the existence of light, and there are few great poets who have not celebrated its presence, or lamented its absence. Long before there was anything that could be called science, the human race recognized that light was a life giver. Nonetheless, it took a very long time for light to begin to be understood scientifically, and even now, some aspects of light remain deeply puzzling.

In 1666, while Newton was formulating the three laws of motion and the universal law of gravity, he was also experimenting with light. Humans had always delighted in the colors of the rainbow at the conclusion of a storm and, by Newton's time, were familiar with the multicolored effects that occurred when light shone through the prisms of chandeliers. Still, people assumed that the light itself was white, and that something about the sky after a storm or the composition of glass added color to it. Newton would later write, "In the year 1666 (at which time I applied myself to the grinding of optick glass or other figures than spherical) I procured me a triangular glass prism, to try the cele-brated phaenomena of colours." (The spellings here are those in use at the time.)

The experiment Newton performed was simple—but no one had ever thought of doing it before. He admitted a narrow beam of light into his workroom by making a small hole in the shutter covering the window. The light was white. Then he placed his prism in front of the beam of light. On the opposite wall, a full spectrum of colors appeared. Newton then took a crucial further step. He used two boards, each with a very small hole in it. He placed one board between the prism and the window, further narrowing the beam of light. The second board was placed between the prism and the wall, so that only a single color was able to pass through the hole in it to appear on the wall. He then put a second prism in front of that hole, and saw that, again, only the single color was reflected on the opposite wall. The second prism did not change the color of the light. He repeated this process for each of the colors of the spectrum, and each time the light that passed through the second prism was unaltered. Thus, the colors were not in the prisms, but in the light itself—other-wise, the second prism should have produced all the colors, not just the isolated one. Light was not white, but contained all the colors of the rainbow, which became visible when the prism split, or refracted, them. It subsequently became clear that the rain-bows that appear in the sky are refracted through droplets of rain, which act as prisms under certain conditions.

Subsequently, Newton performed an additional experiment, recounted in his 1704 book *Opticks,* that used the second prism to recombine the colors and turn them back into white light. With his new understanding of the composite nature of light, Newton tried to solve a problem that affected both microscopes (invented in the Netherlands by Zacharias Janssen in 1609) and telescopes (which Galileo had started constructing, also in 1609, after hear-ing about a refraction lens invented the previous year by the Dutch optician Hans Lippershey). When looking through either of these instruments, a color fringe would appear at the edges, blurring the image being viewed. This problem increased with magnification, making matters worse. In 1668, Newton designed a telescope that made use of a concave mirror, which eliminated the color fringe because the surface of the mirror reflects light rather than breaking it up, or refracting it, the way a lens does. For that reason and because mirrors are both cheaper to manufacture and easier to mount than lenses, most large telescopes today are of the reflecting kind originally devised by Newton.

Newton also suggested that light consisted of what he called "corpuscles," meaning small particles like those in the blood, which were sprayed outward like shotgun pellets. That idea was widely accepted, although the nature of the particles would not be further elucidated for more than 200 years. In the meantime, however, another discovery was made in 1676, by the Danish astronomer Ole Römer. Since ancient times, people had believed that light traveled at infinite speeds, but Römer, while studying the eclipses of Jupiter's moon Io through the telescope at the Paris Observatory, realized that Io did not disappear behind Jupiter at the predicted time. What's more, the observed time would be later when Jupiter was at a greater distance from the Earth and earlier when the giant planet was closer. That meant that light wasn't instantaneously propagated, as scientists had believed since Aristotle first proposed the idea in 350 B.C.; instead, it required a greater time to travel a farther distance. Although Aristotle was a very great figure and was the first to grasp many scientific ideas correctly, his reputation was such that even when he was completely wrong, his views often drowned out revisionist voices. By the seventeenth century, however, his mistaken concept of the solar system had at last been replaced by the work of Copernicus and Galileo, somewhat diminishing his luster. Thus the idea that light traveled at a finite speed was quickly accepted. Indeed, the Danish astronomer's estimation for the speed of light, worked out from his observations of Io, was very close to that in use today: 186,281 miles (298,050 kilometers) per second.

Light was white, though composed of many colors. It traveled at a finite speed, although a very rapid one, nearly a million times as fast as sound. It seemed to consist of particles. All this had been determined by the beginning of the eighteenth century, but things pretty much stalled there for almost another 200 years.

In 1900, German physicist Max Planck published the first paper on what has come to be known as quantum physics, which ran counter to the classic physics of Newton. He established that heated bodies emit energy only in indivisible amounts, which he termed *quanta*. It had previously been assumed that atoms, when "excited" into action, emitted energy in a smooth gradation of values, a curve that could be plotted up or down but was always unbroken. Instead, his experiments convinced him, the energy was broken up into innumerable tiny units, and that each discontinuous quantum contained an amount of energy that was determined by its frequency.

Planck was trying to reconcile work by separate individuals, Wilhelm Wien and Lord Rayleigh (John William Strutt), carried out during the 1890s, which had produced radiation laws that were valid only at high frequencies, in Wien's case, and only at low frequencies in Rayleigh's case. Their work had been based on the assumption that radiation was emitted in waves, but by turning to particles, Planck was able to derive a law that worked at any temperature or frequency. His equation contained a constant, subsequently found to be a fundamental law of nature, and now called Planck's constant. In 1918, he was awarded the Nobel Prize in Physics for this work.

Although the concept of energy consisting of particles was revolutionary, it was quickly seized on by Albert Einstein. In his fourth important paper of 1905, Einstein used Planck's theory to explain the photoelectric effect, proposing that when light particles strike the surface of particular metals, electrons are necessarily jettisoned. The packets of light energy involved (Einstein called them "light quanta," but they were later given the name *photons*) were viewed as particles rather than waves.

In the 1905 paper that laid out the special theory of relativity (expanded into a general theory in 1916) Einstein dealt with a different aspect of light—its speed. His theory held that whether an observer traveled at high speed toward a source of light, or at high speed away from it, the speed of light itself remained the same for both observers. That being the case, however, other things would have to change: In terms of an observer's framework, length would decrease, time would slow down, and mass would increase. At

A surprisingly dapper Albert Einstein is at work in the Swiss Patent Office in Bern in 1905, the year that this obscure young man published his first four papers, which laid the foundation for much of twentieth-century science, and upended the Newtonian universe. Photograph by Lotte Jacobi, courtesy the Lotte Jacobi Archives, University of New Hampshire.

ordinary velocities these effects would not occur, and Newton's laws would continue to rule. But as one got closer to the speed of light, the slowing down of time, for example, would become considerable. If an object, say a spaceship, were to travel at the speed of light, or a greater speed, time on the ship would stop, the ship's length would decrease to zero and its mass would become infinite. Thus, nothing could in fact reach or exceed the speed of light.

Aspects of light that had puzzled Newton more than two centuries earlier had now been explained in ways that revealed a far stranger universe than he could possibly have imagined. Science-fiction writers have long been enthralled, as well as frustrated, by the implications of relativity. On the plus side, the time warping that would take place has figured in many stories and novels that revolve around the idea that a space traveler exploring other galaxies would return as a still-young man or woman, whereas everyone he or she had left behind would have long since died. On the frustrating side, however, all kinds of gizmos, like the engines capable of attaining warp speed on *Star Trek,* have had to be created in order for characters to traipse around the universe at will.

But the new concept of light developed by Einstein has given physicists headaches of their own. Light, like gravity, was supposed to be carried on the ether, and it was in fact experiments concerning the speed of light, carried out by Albert Michelson and Edward Morley in 1889, which resulted in the conclusion that there was no ether—which meant that neither light nor gravity was being transported in that way. This was one of those experiments that did not come out the way it was supposed to. Michelson, an exceptionally brilliant young man who had graduated first in his class from the U.S. Naval Academy at Annapolis four years earlier, and Morley, a very distinguished chemist, fully expected to prove the existence of the ether once and for all. Michelson had constructed an optical interferometer to time the mirrored return of two beams of light fired simultaneously, one into the supposed ether "wind" and one at right angles to it. Because waves have a direction, the ether was supposed to have one, too, and there should have been a difference in the time it took light to move with the wave, as opposed to across it, just as a boat traveling with the wave current in the ocean will move forward more quickly than one traveling athwart the wave. There was no difference at all—none.

The elimination of the ether helped pave the way for Planck, Einstein, and quantum theory in general. Wave theory itself suffered a setback. Perhaps everything was really about particles. Not all physicists were willing to dismiss waves so easily however. Because light reflects off a boundary (or surface) that is capable of reflection, refracts (bends) when it approaches the boundary from an angle, and diffracts (spreads) around the boundary if it is small enough, it is difficult to deny that it is a wave. That is what waves of all kinds do when they hit a boundary: reflect, refract, or diffract. Sound waves do it, ocean waves do it, and so do light waves. This "if it walks like a duck" argument was difficult to refute.

On the other hand, as the technical possibilities for testing quantum theory increased in the course of the twentieth century, leading to the discovery of one new elementary particle after another, many of them predicted to exist years before they were isolated, quantum physics established itself as the most successful theory ever devised. Thus we had a second duck strutting around. The upshot has increasingly been a wide straddle to allow both ducks free passage. This is reflected in the definitions of the word *photon* given, for example, in the 1998 *QPB Science Encyclopedia*: "in physics, the elementary particle or 'package' (quantum) of energy in which light and other forms of electromagnetic radiation are emitted. The particle has both particle and wave properties."

When is it a wave and when is it a particle? Generally speaking, it is regarded as a wave when it is traveling through the vacuum of space, but when it strikes a surface it turns into particles. The wave aspect of light is used by astronomers in determining the shift toward the red end of the spectrum, which in turn is employed to determine how far away a star or galaxy is from Earth. The quantum definition of light is essential to the function of the laser. Many physicists remain deeply unhappy with this split. It prevails largely because it allows a certain amount of "wriggle room": One scientist can say that light is more a matter of waves than it is particles, while another can state that it is more a matter of particles than waves. Depending on the kind of research the scientist is engaged in, both can be "right." This approach is all a bit "squishy" for most physicists, however. Even those on opposite sides of the debate may sometimes wish that the question could be settled once and for all. That would also be helpful to young people studying physics in school or college. Depending on who's in charge of the department, you can be taught in high school that light is primarily a wave, and then have the particle argument elevated to primary importance in college. Some academic Web sites go in opposite directions on the subject.

And well they might, when you get right down to it. As physicist Sidney Perkowitz details in his 1996 book *Empire of Light,* experiments of various kinds have been conducted throughout the twentieth century by some of its most eminent scientists, showing conclusively that light is a wave—and that light is a particle. The structure of the experiment itself appears to change the result, yet by the most rigorous standards of science, the different kinds of experiments are in themselves valid. This echoes the basic quantum paradox that we explore in detail in the next chapter: Electrons (and the photons of light) behave differently, depending on the actions of the observer.

Wave? Particle? Does it even matter? If it can function as both, who are we to quarrel with that? Perhaps the problem is with the scientific obsession about pinning everything down. Sidney Perkowitz is a physicist, but his book about the nature of light is subtitled "A History of Discovery in Science and Art," and he is as interested in art as he is in science. That may make him more comfortable with dualities than most scientists are. Toward the end of his chapter on the subject of the wave/particle debate, he cites Georges Braque, cofounder of cubism with Picasso, who wrote, "The truth exists, only fictions are invented." This, says Perkowitz, is "a guiding precept for those seeking to understand light. Light is what it is. The scientific stories we invent to explain its maddening puzzles only reflect our present ignorance, while reality calmly continues its smooth and true functioning regardless of the tales we tell. And if mind and matter are truly linked, Braque's aphorism may carry a richer meaning; perhaps the truth about light and our fictions simultaneously invent each other."

Some scientific mysteries are annoying because they seem to shut us out of something that it seems we ought to be able to understand—how we learn language, for example, or whether dolphins have one as well. This pairing in itself is doubly irritating because it points up a flaw in ourselves. If we can't figure out how we acquire language, how dare we judge dolphins? Other puzzles are important because if we fail to understand them we may be prevented from certain achievements or, worse, bring harm upon ourselves. The mysteries of how ice ages occur are connected to the potential perils of global warming. Maybe we'd better attain a firmer grasp of how the rise and fall of our planet's temperature occurs—and fast.

The mystery of light can seem more benign than most. In a hundred years time, we moved from the invention of electric light to the harnessing of the power of the laser. The first made it much easier to read a book at night; the second makes it possible to remove cataracts from the eyes with remarkable ease, so that people can read books again, any time of the day. We are doing very well in terms of not merely understanding some of light's mysteries, but also putting that knowledge to work for us. We know how to take advantage of both aspects of its dual nature as wave and particle. Perhaps we can enrich ourselves most in this case by accepting that duality.

Let there be light.

❋ To investigate further

Perkowitz, Sidney. *Empire of Light: A History of Discovery in Science and Art*. New York: Holt, 1996. Of all the books cited in this volume as further reading possibilities, none can be more highly recommended to the general reader than this wonderful small book. It is packed with scientific information, presented in a way that is remarkably easy to absorb. Beyond that, it is a book with a soul.

Feynman, Richard P. *QED: The Strange Theory of Light and Matter*. Princeton, NJ: Princeton University Press, 1985. While not for the casual reader, Feynman is always fascinating, and his puckish sense of humor enlivens some high-level physics.

Westfall, Richard S. *The Life of Isaac Newton*. Cambridge, England: Cambridge University Press, 1993. A scholarly but very readable biography of one of the greatest minds in human history.

Chapter 16

Why Is There So Much Quantum Frustration?

Let's begin with a few words and phrases:

Quark.
Quantum foam.
Quantum tunneling.
Quantum claustrophobia.
Quantum claustrophobia!

The field of quantum physics is littered with cute names. This trend was started by Murray Gell-Mann, who won the 1969 Nobel Prize in Physics for his classification of elementary particles and their interactions. His theory predicted the existence of the *quark,* an elementary particle that is a building block of protons and neutrons, and thus of all matter. Why did he call them quarks? An expert on more subjects than most people could name, Gell-Mann came across a tantalizing phrase from James Joyce's supremely difficult novel *Finnegan's Wake*: "Three quarks for Muster Mark!" Because quarks often come in threes, he thought it was an appropriate name. The quark's existence has since been confirmed in the laboratory, and it has been shown to come in six varieties—up, down, charm, strange, top, and bottom—as well as three "colors"—red, green, and blue—although it should be added that they don't actually have any color in the traditional sense. In fact, no quark has ever been directly observed (they are too small and too elusive), but their existence has been confirmed by elaborate experiments.

Quantum foam is no easier to deal with. Try this definition from Brian Greene's *The Elegant Universe*: "Frothy, writhing, tumultuous character of the spacetime fabric on ultramicroscopic scales." "Spacetime fabric" is the union of space and time, which grows out of Einstein's special relativity; quantum foam has been a big troublemaker for physicists trying to unite relativity and quantum theory.

Quantum tunneling refers to the ability of objects to pass through barriers that Newton's laws say they can't penetrate. This can be said quite simply, but the implications are unsettling, to say the least.

As for *quantum claustrophobia,* you will be glad to know that it is simply another term for quantum fluctuations, which occur because of Heisenberg's uncertainty principle—a concept we will return to because is at the root of many of the difficulties of quantum physics.

Little wonder that Niels Bohr, the Nobel Prize–winning physicist who was one of the fathers of quantum physics, famously stated that anyone who didn't occasionally get dizzy thinking about the subject didn't understand it.

In the end, almost everyone who writes about quantum physics ends up using the word *weird,* for reasons that should by now be obvious. Aspects of it are so bizarre, in fact, that even Einstein, who helped set quantum physics in motion, at one point rebelled against the whole idea. Although the field has developed in ways that leave even the world's foremost scientists muttering to themselves, the degree to which its predictions have turned out to be correct makes it perhaps the most successful theory in the history of science.

Let's go back to the beginning to see how the strangeness grew. In 1900, Max Planck's discovery that the atoms of heated bodies radiated energy in very specific quantities (instead of a continuous stream) gave us the word *quanta* for the particles affected, and the study of such matters was soon designated quantum physics (see chapter 15). Einstein, in 1905, declared that light was made up of particles, or quanta, and these were subsequently designated photons. In 1913, 28-year-old Danish physicist Niels Bohr proposed a structure for the hydrogen atom, which made new use of quantum concepts, providing one of the keys to unlocking the secrets of atoms in general. Einstein's general theory of relativity stole the limelight from quantum physics in 1916, but from 1924 on, there was an enormous burst of new activity on the quantum front.

Prince Louis-Victor de Broglie of France theorized in 1924 that all particles also have a wave function (traveling in waves before becoming particles), turning Einstein's 1905 findings about photons inside out and creating a debate that continues to this day. De Broglie worked out a formula to predict the wavelengths of different kinds of particles, and it was proved in 1927; his success in describing quantum wave mechanics gained him the Nobel Prize in Physics for 1929, one of the quickest such recognitions of a scientific breakthrough. In 1925, 24-year-old German physicist Werner Heisenberg developed the first full-scale theory of quantum mechanics. A few months later, Austrian Erwin Schrödinger came up with a somewhat different approach, using less original mathematics, but he was soon able to show that his and his German rival's model were in fact equivalent. They also had the same problem: What were these waves, anyway? Like the famous scene from the movie *Chinatown* in which Faye Dunaway keeps alternating between saying, "I'm her sister," and "I'm her mother" as Jack Nicholson slaps her face, photons of light, for example, seemed to change their minds about what they were every time they got hit: "I'm a particle"; "I'm a wave"; "I'm a particle"; "I'm a wave." The Dunaway character, for the uninitiated, had been impregnated by her own father; many physicists found the behavior of photons and other subatomic particles almost as horrifying.

An explanation of this duality was provided by German physicist Max Born. The wave aspect of a particle was a description of the *probability* that it would develop a specific characteristic—a position, say, at a certain time. Waves could be divided up into halves or thirds and could even become superimposed on one another, but you couldn't have half an electron. Thus, the waves were the way an electron preserved a fractional possibility of alternative futures as a particle. This was too much for Einstein. He wrote to Born in 1926 saying, "I shall never believe that God plays dice with the world." (Born happens to be the maternal grandfather of the Australian pop star Olivia Newton-John, which does suggest an odd throw of the dice but is in no way "strange" in the quark sense.)

Partially because of Einstein's outrage, Born would have to wait until 1954 for his Nobel Prize, but Werner Heisenberg would get one in 1932 for his 1927 development of the uncertainty principle, which remains to this day the core of quantum physics. It states, with alarming straightforwardness, that it is impossible to know both the position and the speed of a subatomic particle at the same time, because the very act of measuring such a particle will change either its position or its speed. We have all encountered the uncertainty principle in daily life. When we use a ruler to measure a picture we want to frame, for example, we often give the picture itself a little accidental nudge and have to pull it back into alignment. This kind of thing really doesn't matter in the large-size world we live in. At the subatomic level, however, a far smaller nudge will send an electron bouncing off at a great rate. Even the photons in a beam of light are enough to change the nature of a subatomic system. In addition, the greater the precision of one measurement, for position, say, the greater the disturbance to the speed. Subatomic particles *will not allow themselves to be pinned down*. This uncertainty, surprisingly, can be useful. Quantum tunneling is one result—because there is always the possibility that a subatomic particle, by changing its nature for a *nanosecond* (one billionth of a second), will go through a barrier that it should not be able to penetrate. The probability of that happening can be harnessed—and has been, in the "scanning tunneling microscope," originally created by Gerd Binnig and Heinrich Rohrer at the IBM research center in Zurich, Switzerland, in 1981 and now in widespread use. The STM, as it is called for short, can be used to reveal the surface of an object in such detail that rows of atoms a billionth of a meter apart can be photographed.

Werner Heisenberg is shown here as a young man, around the time he formulated the "uncertainty principle," which defines both the possibilities and the essential weirdness of quantum physics. Courtesy the American Institute of Physics, Emilio Segré Visual Archives, Segré Collection.

To return to the 1920s, when a major paper on quantum physics seemed to appear every week, Austrian physicist Wolfgang Pauli declared in 1927 that no two particles within an atom could have the same quantum numbers (which is why there are different varieties and colors of quarks). Pauli's exclusion principle was easier to grasp than many quantum discoveries, but it had a particularly large impact because it extended the reach of quantum theory to another branch of science. The periodic table of elements, devised in the nineteenth century by Russian chemist Dmitry Mendeleyev and subsequently added to by others, classified the elements by their atomic weights. The table was periodic in the sense that similar elements, such as sodium and potassium, appeared at predictable intervals. No one had figured out why that should be so, but Pauli's exclusion principle explained the problem.

It was known that atoms were surrounded by electrons orbiting it much as the planets do the Sun. Now, it became clearer how this system worked. As Curt Suplee puts it in *Physics in the Twentieth Century,* "As atoms got larger, they filled each successive energy level or 'shell' of electrons until adding another would put two electrons in the same quantum condition. At that point, the electron had to go into the next shell. The number of electrons in the outermost unfilled shell determined the element's reactive properties. Chemistry had become a quantum affair." In 1931, Pauli also predicted the existence of the *neutrino* (a particle without a charge, consisting of three quarks, found in the nucleus of an atom). The existence of the neutrino was not confirmed until 1955, but Pauli was awarded the 1945 Nobel Prize in Physics for his discovery of the exclusion principle. He had to wait longer than most of the quantum pioneers to be rewarded, perhaps because he was an irascible man with a wicked knack for putting down other scientists. His response when presented with a subpar idea: "It's not even wrong."

An analysis of electrons that accounted for their *spin,* another new unit of quantum measurement, was made by 23-year-old English physicist Paul Dirac. In the process he discovered something else that further stunned the already reeling scientific world of 1928. To Dirac's own initial dismay, it appeared that there had to be a counterpart to each electron with a "negative energy." This first suggestion of the existence of *antimatter* (which would destroy matter if the two collided) was so unnerving as to create doubt in the minds of some concerning Dirac's other findings. A mere four years later, however, American physicist Carl Anderson, studying cosmic rays with the new cloud chamber at the California Institute of Technology, found particles that were both negatively and positively charged. It is easy to get confused here because in its "normal" state the electron has a negative charge. Thus an electron with a positive charge would be an "antielectron," which in terms of the English language has negative connotations. These antielectrons were later given the name *positrons*. Anderson's experimental production of positrons came as a great relief to Dirac—he had not made some kind of mistake but instead postulated, correctly, the existence of the sci-fi-like antimatter. Dirac would share the 1933 Nobel Prize in Physics with Erwin Schrödinger, and Carl Anderson would share the 1936 prize with another cosmic-ray researcher, Victor Hess of Austria.

As ever greater knowledge of the quantum realm was attained, fissures began to appear in the field. Things were getting more bizarre by the month, and Einstein was hardly alone in objecting to some of these peculiarities. Although Einstein himself had dethroned the Newtonian universe, at least on the large scale, he was still a classicist, and he found it deeply disturbing that quantum theory and relativity were not meshing better, and that quantum mechanics seemed to overturn Newton's work not merely on relativistic scales but also in terms of the observed reality of everyday life. He and Niels Bohr, who were good friends with enormous respect for one another, spent years debating these issues.

Bohr came up with a way to try to bridge the gap between quantum theory and the rest of physics. His solution (known as the "Copenhagen interpretation" because that was where he worked) claimed that particles always had the properties of waves until they were acted on by being observed, at which point, because of the observer's presence, they could become particles. In other words, the quanta remained in an indeterminate wavelike state (in agreement with Heisenberg's uncertainty principle) until an observer took a look at what was going on. The act of observation itself, of taking a measurement, would "collapse the wave function," with its inherent multiple possibilities, and cause the particle to resolve itself into one of its potential states.

While the Copenhagen interpretation satisfied most physicists, and still does, it has met resistance from some of the greatest scientific minds of the century. Erwin Schrödinger, a year after winning the 1933 Nobel Prize, conceived of a *thought experiment* (a purely intellectual exercise with the logic of a laboratory experiment)

Erwin Schrödinger, photographed here, shared the 1933 Nobel Prize in Physics with Paul Dirac for their development of quantum physics. Two years later, Schrödinger challenged the "Copenhagen interpretation" of Niels Bohr with his famous thought experiment concerning a cat that was both dead and alive. Courtesy the American Institute of Physics, Emilio Segré Visual Archives.

intended to show the absurdity of the Copenhagen interpretation, and it has survived as perhaps the most famous thought experiment in the history of science. Imagine a box in which a live cat is placed. A container of radium is put into the box also, as well as a vial of cyanide gas. Radium is subject to atomic decay. If the radium decays during the hour the cat is left in the closed box, it will trigger the release of the cyanide from the vial, which will kill the cat. If it doesn't decay, the vial will remain unbroken and the cat will be alive. According to the Copenhagen interpretation, however, until the box is opened and the result observed, the cat in the box is both dead and alive simultaneously because both probabilities exist. Moreover, it will continue to be dead/alive until someone looks in the box, at which point the uncertainty will be at an end, and the cat will either be thoroughly dead or thoroughly alive.

The implications of this thought experiment can be made clearer by extending it to an everyday situation. A businessman known to have high blood pressure travels to Cleveland and stays at a hotel, preordering the 8:00 A.M. delivery of breakfast to his room. Once he closes the door to his room, this man with high blood pressure is both alive and dead until the waiter arrives in the morning. If the businessman opens the door to the knock, the multiple probabilities are resolved, and he is still alive. If he doesn't answer the door, the waiter will report a problem, a key will be used, and the man with high blood pressure will be found dead in his bed of a heart attack. (The possibility that he was in the shower because his watch is slow doesn't count.) Obviously, this is absurd—which was just what Schrödinger was trying to show. To this day, "Schrödinger's cat" seriously annoys physicists who defend the Copenhagen interpretation. Stephen Hawking says that when he hears it mentioned, "I reach for my gun."

The fact that this thought experiment still hits raw nerves suggests why it remains so powerful: Some problems with quantum physics are very difficult to explain away. Einstein later conducted a thought experiment in conjunction with two other physicists— Boris Poldosky and Nathan Rosen. It had been shown in laboratory experiments that two electrons shot through different holes during laboratory experiments somehow "know" what has happened to the other. Einstein and his colleagues amplified the distance involved to a light-year of separation in space. (This kind of mathematical amplification is often used in physics to make matters clearer.) The result is that two electrons a light-year apart must be able to instantaneously communicate with one another, using some kind of signal that travels faster than light—an impossibility in terms of relativity. "Spooky action at a distance," Einstein said with displeasure.

Quantum mechanics *work*—lasers wouldn't function otherwise, and scientists have already shown in the laboratory that it should be possible (if the problems of scale can be solved) to build a quantum computer that will make use of the fact that electrons know what one another are doing at a distance. The how and why remain deeply mysterious, however. Gravity has not yet been successfully incorporated into quantum theory—and gravity certainly works, too. Many scientists have reached a point where they shrug their shoulders and say the why doesn't matter that much anymore, let's just apply the knowledge generated by quantum mechanics.

That resolution isn't good enough for other physicists. They want to know not only why it works but exactly where the line should be drawn between the quantum realm and the everyday Newtonian reality. At what scale does the probability that governs the quanta turn into the "decision" that makes a cat exist—even the corpse of a cat? It has now been established that subatomic particles wink in and out of existence constantly in what amounts to no time at all. Where are they coming from, and where are they going? Some scientists clearly wish that Schrö-dinger's cat not only were dead, but also would vanish from sight the way corpses sometimes do in horror movies.

Since the mid-1990s, quantum physics has reached another level entirely, as exemplified by superstring theory, which we discuss in chapter 20. That, too, has both pluses and minuses attached to it. Some scientists also wonder whether such grand attempts at theories of everything aren't getting a bit ahead of things. That cat from 1935, simultaneously dead and alive, has never been given a proper burial.

⚛ To investigate further

Gribbin, John. *In Search of Schrödinger's Cat.* New York: Bantam, 1984. This was one of the first books to try to convey the peculiarities of quantum theory to a popular audience and is still well worth reading.

Gribbin, John. *Schrödinger's Kittens and the Search for Reality.* Boston: Little, Brown, 1995. This successor to Gribbin's 1984 book is well-written and lucid but tends to be uncritical about concepts that some physicists have seriously challenged.

Suplee, Curt. *Physics in the 20th Century*. New York: Abrams, in association with the American Physical Society and the American Institute of Physics, 1999. Although the text takes second place to the photographs in this book, it gives a very clear timeline concerning the development of quantum physics and provides considerably more clarity than most such efforts.

Lindley, David. *The End of Physics*. New York: Basic Books, 1993. This book provides a critical analysis of the directions physics began to take in the 1980s toward ever-less-testable theories. Its skepticism was endorsed by 1988 Nobel Prize winner Mel Schwartz.

Perkowitz, Sidney. *Universal Foam: From Cappuchino to the Cosmos*. New York: Walker, 2000. As in his book on light (see chapter 15), Perkowitz approaches quantum physics with an eye to the marvelous connections between our everyday reality and theoretical science. The result is a splendid mixture of clarity and charm.

Frayn, Michael. *Copenhagen*. New York: Anchor, 2000. This internationally acclaimed, Tony-winning play focuses on a meeting between Niels Bohr and Werner Heisenberg in Copenhagen during World War II, when Heisenberg had been asked by Hitler to develop an atomic bomb. The meeting did in fact take place, but what was discussed is unknown. Frayn has imagined the possibilities inherent in the situation in a way that relates quantum physics to the "uncertainty principle" that also underlies human interactions.

Chapter 17

What Are Black Holes Really Like?

Did J. Robert Oppenheimer fall into a black hole? From a number of recent books on cosmology, it might seem so. His name appears nowhere in the indexes of these books, and their lengthy discussions of the complex theories and mathematics surrounding black holes do not mention him. Yet it was this great American physicist— famous to this day as head of the Los Alamos team that built the atomic bombs dropped on Hiroshima and Nagasaki—who first conceived of these deeply strange cosmic entities as an inevitable implication of Einstein's theory of relativity. At the end of 1938, Oppenheimer, working together with George Volkoff, had completed a computation of both the masses and circumferences of neutron stars. This work had convinced Oppenheimer that massive stars had to implode when they died. What, he asked, would the results of that implosion be?

Oppenheimer enlisted the help of a brilliant and independent-minded graduate student of his at the California Institute of Technology, Hartland Snyder, to work on the mathematical equations involved. Kip Thorne, one of the world's foremost current experts on black holes, does discuss Oppenheimer's work in detail in his 1994 book *Black Holes and Time Warps*, even though, ironically, he was a student of one of Oppenheimer's great rivals and oft-time

J. Robert Oppenheimer was the first to propose the theoretical existence of black holes, in 1939. Despite the mathematical strength of his arguments, the concept was initially resisted by most physicists, and his work on the subject was put aside when he was named to head the team of scientists developing the atomic bomb at Los Alamos. Courtesy the American Institute of Physics, Emilio Segré Visual Archives.

critics, John A. Wheeler. Thorne notes that the calculations undertaken by Snyder, with the guidance of not only Oppenheimer but also Richard Tolman, were formidably difficult. Aspects of the problems involved would not be solved until the advent of supercomputers in the 1980s. "To make any progress at all," Thorne has written, "it was necessary to build an idealized model of the imploding star and then compute the predictions of the laws of physics for that model." Snyder, in what Thorne terms a tour de force, set up the applicable equations and solved them. "By scrutinizing those formulas, first from one direction and then another, physicists could read off whatever aspect of the implosion they wished—how it looks from outside the star, how it looks from the inside, how it looks from the star's surface, and so forth."

Many physicists found what these equations showed to be almost incomprehensible. The problem was that from the external frame of reference the implosion would reach a point where it froze forever. To an observer on the star's surface, however, being carried inward with the implosion, it would not seem to freeze at all. The idea that a star could give the appearance of doing two entirely different things simultaneously depending on one's vantage point constituted a warping of time beyond any previously considered. Yes, Einstein had showed that time did warp. Yes, quantum theory and Heisenberg's uncertainty principle suggested that the very act of observing could alter what was happening—but that was at the subatomic level. This was carrying things too far in the view of most American physicists.

In fact, the 1939 Oppenheimer/Snyder paper had some precursors. Eleven years earlier, young physicist Subrahmanyan Chandrasekhar had theorized that stellar cores more than 1.4 times the size of our sun could not become the often observed white dwarfs; instead, they would continue to collapse because of their gravity. Lev Davidovich Landau, a legendary Russian physicist, came to the same conclusion more or less simultaneously, and he and Chandrasekhar would share the Nobel Prize in Physics for 1983 for their initial work on this subject. Note the time gap here. When scientists have to wait 55 years for a Nobel, it means that their work was way ahead of its time. In 1928, one of the giants of physics, Sir Arthur Eddington, whose measurements during the 1919 solar eclipse had confirmed the warping of space predicted by Einstein's theory of relativity, was outraged by Chandrasekhar's theory. "I think there should be a law of nature to prevent a star from behaving in this absurd way!" he exclaimed.

The Oppenheimer/Snyder paper met with almost the same reaction from John A. Wheeler and other American scientists. There the matter rested for some time, due to the start of World

War II. American physicists were caught up in the practical difficulties of building an atom bomb. After the war, the differences between Oppenheimer and Wheeler, which had begun to reach a personal level (they were both at Princeton's Institute of Advanced Study by now), came to a new head when Oppenheimer initially opposed the development of the hydrogen bomb on both practical and ethical grounds. He was eventually won over on the practical issues but was never happy with the ethical situation. Wheeler, on the other hand, was one of the chief architects of the hydrogen bomb. Oppenheimer's opposition to this new weapon cost him dearly in the McCarthy-ite 1950s, when his security clearance was revoked. Although disloyalty was never substantiated, that cloud may have something to do with the fact that he is neglected in so many discussions of black holes. Another reason for this neglect was that Wheeler did a complete turnabout on the subject.

Indeed, Wheeler's conversion was so complete that he would give them the name *black holes* in 1969, and in becoming one of the most important theorists on the subject, he eclipsed his old rival Oppenheimer quite thoroughly. It will interest *Star Trek* fans that an early episode made reference to these phenomena in 1967. The author of *The Physics of Star Trek,* Laurence M. Krauss, wrote, "When I watched this episode early in the preparation for this book, I found it amusing that the 'Star Trek' writers had gotten the name wrong. Now I realize that they very nearly invented it!" The series writers had used the term "black star."

The early *Star Trek* reference points up the degree to which the general public came to be fascinated by the concept of black holes. In part, that fascination may be due to John Wheeler's name for them, which manages to evoke vast mysteries even as it lends itself to a wide range of bad jokes about the problems of daily life. The wide public has never had much interest in many other important kinds of stars, from white and brown dwarfs to neutron stars, but black holes have demonstrated the kind of hold on the imagination that comets have had for centuries. This is especially odd, in that the world's foremost physicists have been tearing their hair out about black holes for 60 years, and they continue to do so. It has been suggested numerous times, in fact, that the reason for the public fascination with black holes is that the very difficulty in explaining them turns them into a kind of exotic blank slate, on which individuals are free to write what they choose.

Most glossary definitions of black holes center on the idea that their gravitational field is so intense that nothing, even light, can escape them. Kip Thorne carries that further. Although his book on the subject was published in 1994, several years before astronomers began to pinpoint their actual existence, he was on the cutting edge of theoretical developments in regard to them, and his definition adds another wrinkle: "An object (created by the implosion of a star) down which things can fall but out of which nothing can ever escape." Even he is being cautious, however, because his own discussion of black holes leads to stranger conclusions yet.

At this point, let's ask a simple question: How big is a black hole?

Theoretically, anything can become a black hole. A star, a moon, the Empire State Building, an elephant, you, me, a paperweight—if enough force is brought to bear on an object and compresses it to the point that its gravitational field is strong enough to bend space and prevent light from escaping, it will have become a black hole. You and I, even if we are considerably overweight, would end up as a very small black hole indeed, billions of times smaller than an electron. If the Earth became a black hole, its radius would be compressed to that of less than a Ping-Pong ball. The Sun's radius as a black hole would be about a mile and a half (2.4 kilometers).

Realistically, the Sun is not going to become a black hole, never mind you and me. None of us is anywhere big enough to start with. Some stars are big enough, however, so big that they will inevitably become black holes. As Timothy Ferris explains in *The Whole Shebang,* "Every healthy star represents a balance between two opposing forces. Gravity tends to collapse the star. Heat generated at the core radiates outward; its tendency is to blow the star apart. Caught in the balance stars pulsate a bit, owing to the teeter-totter of inward pulling gravity and outward-pushing radiative heat. The pulses are modulated by an elegant feedback mechanism." That feedback mechanism between heat and gravity can keep a star burning for a very long time, about 10 billion years in the case of our sun, which is halfway through its life cycle. The nuclear fuel at the core of a star, which is essential to the feedback mechanism, is burned at a rate that increases by the cube of the star's mass. Thus a star 10 times as massive as our sun would use its fuel a thousand times faster, burning very brightly but far less long. For a star of any size, once the equilibrium between heat and gravity starts to fail, collapse is inevitable.

Stars the size of our sun, or having up to 1.4 times its mass, will become white-dwarf stars, about the size of the Earth but with the mass of our sun; they won't collapse further because of a rule of quantum mechanics, known as the Pauli exclusionary principle (mentioned in chapter 16), which keeps electrons flowing in a way that limits the density of the star. Larger stars will collapse even further, to a diameter that usually is less than 10 miles (16.1 kilometers); these are called *neutron stars* because their core consists largely of these electronically neutral subatomic particles. A neutron star can rotate as many as a thousand times per second, and if they have a magnetic field, they will produce intense, beeping radio beams, which have led to their being called *pulsars.*

Still larger stars may have such an enormous mass that the different kinds of conditions that prevent a white dwarf or a neutron star from collapsing still further are overwhelmed, and a black hole will come into being. Because nothing, not even light, can escape the gravity of a black hole, anything that gets near enough to it to cross its *event horizon* will be sucked into it—that is the point in space at which the normal gravitational rules of the universe cease to operate and those of the black hole take over. The black hole is thus a singularity, a zone within which unique laws apply. There have been many different theoretical attempts to define what goes on inside a black hole. Even Hollywood has had a go at it, in the visually spectacular but dramatically silly Disney film of 1979, *The Black Hole.* Some cosmologists have suggested that anything that fell into a black hole would be stretched into spaghetti-like strands, while others have envisioned the possibility of traveling through a black hole into a different universe. Great minds and innumerable equations have been devoted to such scenarios, but the plain fact is that nobody really knows what would happen. As with some aspects of the Big Bang theory of the universe, the very fact that you are dealing with something that is a singularity tends to create a certain license in describing it. No matter how elegant the mathematics, it is still an imagined reality that is being promulgated.

Since John Wheeler changed his mind and endorsed the concept of black holes, numerous major cosmologists have attempted to nail down the nature of these bizarre stellar entities. Throughout the 1970s and 1980s, and on into the 1990s, theories were as plentiful as the arguments they aroused. Despite this abundance of theories, there was a problem: The actual existence of a black hole had never been confirmed.

There is a built-in problem for astronomers in dealing with black holes. By definition, they cannot be seen. They can only be inferred from what is happening to other stars and galaxies around them. With the repairs to the flawed Hubble Space Telescope in a 1994 space walk, however, and the further development of X-ray telescopes, actual observations that would provide the information on which to base such inferences began to accumulate. The latter part of the 1990s and the beginning of the year 2000 confirmed that many predictions concerning black holes were right in line with the data being recorded. Over the past few years, almost all cosmologists have come to the conclusion that we now have evidentiary proof of the existence of black holes. As is often the case when new information finally starts pouring in, the results have raised as many questions as they have answered.

Astronomers had been quite sure what they ought to be looking for since 1974, when the star Cygnus X-1 (Cyg X-1) first became generally accepted as the best candidate yet discovered for the designation of black hole. Cyg X-1 was a *binary system*—a pairing of two stars quite common in the universe—but a binary of a particular kind: One star was bright in terms of optical viewing, but dark in terms of X-ray measurements, and appeared to be revolving around another star that was the opposite—optically dark but "bright" when observed by X-ray astronomy. Using mathematical formulas developed to determine the weight of stars, it was clear that the dark companion was too heavy to be a neutron star. With such a large mass, it was strongly suspected of being a black hole. By the mid-1980s, astronomers around the world amassed evidence about Cyg X-1, leading Kip Thorne to bet Stephen Hawking that it was indeed a black hole: If Thorne was eventually proved right, Hawking was to buy him a subscription to *Penthouse,* while Thorne was to give Hawking a subscription to the satirical magazine *Private Eye* if matters proved otherwise. Additional evidence had made Thorne 95% sure he was correct by 1990, but he did not expect Hawking to concede. However, as Thorne writes, "Late one night in June 1990, while I was in Moscow working on research with Soviet colleagues, Stephen and an entourage of family, nurses and friends broke into my office at Caltech, found the framed bet, and wrote a concessionary note on it with validation by Stephen's thumbprint."

Cyg X-1 is among the black holes to be confirmed by the combination of optical evidence from the Hubble and new X-ray observations. But other new information has been more provocative. As some astronomers had predicted, observational evidence in the late 1990s produced more and more data to suggest that there were two different kinds of black holes. Scientists were finding not only black holes with the masses of typical binary stars like Cyg X-1, but also black holes with masses equivalent to billions of suns. What's more, these supermassive black holes were found again and again to lie at the center of galaxies—more than 30 of them had been identified by 2001 by measuring the velocities of disks of gas trapped by the black holes, swirling around them like water around a drainhole after a storm.

The findings showed that the bigger the galaxy, the bigger was the black hole at its center. In addition, the supermassive black holes seemed to exist only in galaxies of an elliptical shape, which had a dense bulge of stars at the center. Galaxies with no central bulge appeared to lack black holes altogether. Our own Milky Way galaxy, which has a relatively small central bulge, has black holes, but only the smaller-sized holes with masses equivalent to a few suns. Whether the black hole is very large or relatively small, its mass in relation to the central bulge of the galaxy in which it is found always amounts to 0.2%.

Cosmologists examining this evidence are becoming increasingly convinced that black holes may be the seeds around which galaxies themselves form. After a team discovered three additional supermassive black holes, its leader, Douglas Richstone of the University of Michigan, said in January 2000, "Somehow, these black holes, when they determine their mass, they know the mass of the galaxy they're sitting in, or when the galaxy is forming, it knows the mass of a black hole that it is forming around or that it appears in. These are mutually regulated in some way." It has long been recognized that at the quantum level, electrons can "know" what one another are doing, but that such mutuality could occur on the galactic scale mystifies and excites cosmologists in equal measure. Currently, there is considerable chicken-or-the-egg debate about which came first, the galaxy or the black hole. Some scientists think that the black hole did, but others believe that their development may be completely intertwined.

Back in 1939, when the Oppenheimer/Snyder paper was published, suggesting the existence of black holes, it was derided by some of the most prominent cosmologists in the world. Gradually, most scientists were won over to the view that black holes had to exist, but it was not until the late 1990s that the Hubble telescope would make it possible to clearly view the galactic perturbations that confirmed their presence in galaxy after galaxy. Still, black holes have only begun to yield up their secrets, and they have created new mysteries which, while they may hold the key to vastly increased understanding of how the universe works, seem likely to create as many complications as solutions for a long time to come.

⚛ To investigate further

Thorne, Kip S. *Black Holes and Time Warps: Einstein's Outrageous Legacy*. New York: Norton, 1994. Although this book was published before the Hubble telescope was functioning properly, and it is therefore out-of-date in some respects, it remains the best account by far of the development of black-hole theory. While weighing in at 600 pages, it is remarkably readable and clear and is enlivened greatly by Thorne's insider stories about the personalities of the scientists who have pushed forward the boundaries of black-hole study over the preceding 60 years.

Pickover, Clifford A. *Black Holes: A Traveler's Guide.* New York: John Wiley & Sons, 1996. This very popular and well-received book takes the reader on an imaginary journey with two scientists of the future as they travel to a black hole to carry out a series of experiments. It is a lively approach to a difficult subject, with terrific illustrations and even codes that allow the reader to create black-hole computer graphics on a personal computer.

Couper, Heather (with contributions by Nigel Henbest and illustrations by Luciano Corbella). *Black Holes.* New York: DK Publishing, 1996. This is a "young adult" book, but a sophisticated one, and it may be just the thing for readers who want to know more about black holes but do not have the time or inclination to deal with a massive work such as Kip Thorne's.

Ferris, Timothy. *The Whole Shebang.* New York: Simon & Schuster, 1997. As with the many other cosmological subjects covered in this wide-ranging book, Ferris does a good job of explaining complex material.

Wheeler, John Archibald, and Kenneth William Ford. *Geons, Black Holes, and Quantum Foam: A Life in Physics.* New York: Norton, 1998. Wheeler has been at the center of quantum physics since the late 1930s, and his autobiography provides an insider's view of some of the most extraordinary cosmological developments of the twentieth century.

Chapter 18

How Old Is the Universe?

n 1912, at the cluttered offices of the Harvard College Observatory in Cambridge, Massachusetts, a discovery was made that would completely reshape the field of astronomy. Its effects are still being felt today and lie at the center of major arguments concerning the size, shape, age, and ultimate fate of the universe. Most confounding to astronomers is the question of the universe's age. Measurements taken by different teams of equally eminent scientists not only suggest answers that are billions of years apart, but even worse, keep running into the same impossibility: a universe that is younger than the oldest stars it contains.

The 1912 discovery was not made by a prestigious astronomer. Henrietta Swan Leavitt was one of a group of women who worked at the observatory offices sorting and categorizing photographic plates of the stars that were made using the Harvard College telescope in the mountains of Peru. Despite the fame of Marie Curie, who had shared the Nobel Prize in Physics in 1903 and been the sole recipient of the Nobel Prize in Chemistry in 1911, women scientists were rare at the time. The work done by Leavitt was crucial—she and her coworkers were fondly known as "the computers"—but it was also tedious and not very well paid. Nevertheless, as she studied a series of plates taken of the Magellanic Clouds, Leavitt realized that differences in the brightness of what were known as Cepheid stars had to be the result not just of their size but also of their distance from the Earth.

Spiral galaxy NGC 4639 is located 78 million light-years from Earth in the Virgo cluster of galaxies. The bright spots around the periphery are young stars, some of them variable Cepheid stars, the significance of which was first recognized by Henrietta Swan Leavitt in 1912. Cepheids have been used since the 1920s to measure stellar distances. Courtesy NASA (A. Sandage, Carnegie Observatories; A. Saha, Space Telescope Science Institute; G. A. Tammann and L. Labhart, Astronomical Institute, University of Basel; F. D. Macchetto and N. Panagia, Space Telescope Science Institute/European Space Agency).

The importance of this observation was quickly recognized by American astronomer Harlow Shapley, who would later head the Harvard College Observatory from 1920 to 1952. Cepheid stars have an unusual characteristic. Each of them waxes and wanes in brightness over a period ranging from a few days to a few weeks, in an endlessly repeated cycle. By observing as few as two cycles, it is possible to determine a specific value for the brightness of the star, called its *absolute magnitude.* The difference between that value and how bright it seems to be—its *apparent magnitude*—is a factor of its distance from the Earth. Isaac Newton had established that the brightness of an object lessens according to the square of its distance from the viewer. The distance can be calculated using basic trigonometry, a formula long employed by sailors, for example, to determine the distance from a ship to a lighthouse. In astronomy, the ship is the Earth, and the lighthouse is the Cepheid star.

Using this new tool, Shapley made further studies of the Magellanic Clouds, and in 1916, he announced that our own solar system lay on the outskirts of the Milky Way instead of near its center, as astronomers had long assumed. He estimated that the real center was 50,000 light-years away. This figure was later corrected to 30,000 light-years. Shapley got the figure wrong because he continued to give credence to another assumption—that the entire universe was contained within the Milky Way. Even when making great breakthroughs, scientists quite often fail to see the whole picture because they cling to one traditional view while exploding another.

It would be Shapley's chief rival, Edwin Powell Hubble, who would make the even more startling assertion that the Milky Way was just one piece of a much larger puzzle. In a paper read for him by the leading astronomer Henry Norris Russell on New Year's Day, 1925, at an important meeting of astronomers in Washington D.C., Hubble showed that the Milky Way was just one galaxy in a great sea of space containing innumerable other galaxies. He called them "island universes," a phrase both poetic and apt, which made clear even to the ordinary person the vastness of the cosmos he had newly disclosed.

Hubble, along with a few other mavericks, had believed that the spiral nebulae were not mere clouds of swirling gas within the Milky Way, but entire star systems far beyond its confines. As an associate astronomer at the Mount Wilson Observatory in California, he had used the new 100-inch (2.5 meter) telescope put into service in 1923 to gather photographic evidence of his theory. It was backed by calculations using the Cepheid stars in the spiral nebulae, revealed for the first time to human view by the new telescope. About 30% of known galaxies are now classified as *spiral nebulae*. They consist of a central bulge of stars and a flattened disc that usually contains two spiral-shaped arms, made up of hot young stars, as well as clouds of dust and gas. The extent of the universe uncovered by Edwin Hubble left astronomers agog, and at first, it seemed almost incomprehensible to the general public.

Throughout human history, nothing has been quite so efficient at undermining our self-importance as astronomy. Good old Ptolemy made us the center of everything back in the second century A.D., by constructing a universe in which the Sun, the other planets, and all the stars revolved around the Earth itself. Humans were so pleased with that picture that it survived until the sixteenth century, when Copernicus showed that the Earth revolved around the Sun. This debasing concept was fiercely resisted for almost a century; as late as 1633, Galileo got hauled before the Inquisition for championing it. By the end of the first two decades of the twentieth century, with Shapley moving our solar system to the suburbs of the Milky Way, and Hubble showing that there were numerous other galaxies, the jig was really up. We would have to get used to the idea that we live on a minor planet of a minor star in one of several hundred million galaxies, many of which contain more than 2 billion stars.

After unleashing his 1925 bombshell, Hubble went back to work studying the redshift in Cepheid stars in the spiral nebulae he had identified as galaxies. *Redshift,* a change in color toward the red end of the spectrum, occurs when the source of light is moving away from the viewer. This phenomenon had been studied by astronomer Vesto Slipher at the Lowell Observatory in Flagstaff, Arizona, for a number of years, but Slipher had moved on to other matters by 1922, and it was Hubble who reached the conclusion that redshift indicated that other galaxies were moving outward, expanding the size of the universe as they went. Hubble's Law, published in 1929 and still a basic tool for measuring the size and age of the universe, held that the farther away a galaxy is, the greater the redshift that will be exhibited by its spectrum.

At that time, American astronomers were sharply focused on the observations made through the telescopes at Mount Wilson and the Lowell Observatory, which were far superior to any found in Europe. European physicists, on the other hand, led by Albert Einstein, were using mathematical theories to describe the universe. By the 1930s, the astronomers and the physicists began to recognize that they were approaching the same problems from different points of view, and a greater meshing of theory and observation began to occur. Out of this cross-fertilization, the Big Bang theory was born. As described in detail in chapter 1, this explanation of cosmic origins holds that 10 to 20 billion years ago all the matter and energy in the universe was concentrated in one point, infinitely dense and hot, which suddenly exploded; the matter and energy released then evolved into the vast galaxies we know today.

As we have seen, the Big Bang theory was not taken very seriously until the existence of a theoretically postulated cosmic microwave background (the haze left over from the Big Bang) was confirmed in the 1960s. The efforts of astronomers and physicists over the course of the twentieth century, from Einstein's original paper on relativity in 1905, through Henrietta Leavitt's recognition of the importance of Cepheid stars and Hubble's application of that work in identifying the multitude of island universes—all were coming together with the contemporary discoveries of radio astronomy to provide a real basis for determining the size, age, and fate of the universe.

Then came the Hubble Space Telescope, appropriately named for the man who had first shown that there were innumerable galaxies. It was expected that the Hubble telescope would confirm the general view that the universe was between 14 and 20 billion years old. Earthbound telescopes can detect Cepheid stars as distant as 15 million light-years away. Once the Hubble was fully operational, following the repair of its defective main mirror during a 1993 space walk, it became possible to view Cepheids at a distance of 60 million light-years.

The first report from a team of astronomers using data from the Hubble, published in 1994, threw everything into utter turmoil. It had been widely accepted that the *Hubble constant* (the rate at which the universe was expanding, in keeping with Edwin Hubble's 1929 Law) equaled 50 kilometers per second per megaparsec. Fifty kilometers (31 miles) is a number anyone can easily grasp, but a megaparsec is a number of an altogether different magnitude. A single parsec is 3.26 light-years. A megaparsec is 1 million times that. In a universe in which our nearest neighbor among the galaxies, Andromeda, is 2 million light-years away, astronomers take for granted such extraordinary numbers. They do not like it, however, when new observations cause the numbers to change drastically—and that was what happened in 1994.

The 22-member team using the Hubble telescope had studied 20 Cepheid stars in the galaxy M100, in the core of the Virgo supercluster. The redshift of these Cepheids led the team to the conclusion that M100 was much closer to us than had been thought—so much closer, in fact, that it raised the number of the Hubble constant from 50 to 80 kilometers (31 to 50 miles) per second per megaparsec. That meant that the universe was expanding at a much faster rate than previously believed. If it was expanding that fast, then it had to be younger—a lot younger. Instead of being 14 to 20 billion years old, it now seemed to have an age of only about 8 billion.

That was not just a number that was difficult to swallow, but one to choke on. The oldest stars in our own Milky Way galaxy, which had been studied the longest and most thoroughly, had been determined to be around 14 billion years old. That would make them older than the universe as a whole, an obvious impossibility.

In the ensuing panic, some astrophyicists even suggested reviving Einstein's cosmological constant, the antigravity force he had used as a fudge factor in developing his theory of relativity and had then discarded. Clearly, it was easier to assume that something had gone wrong with this one study using the Hubble telescope—however eminent those associated with it might be. The team went back to work. In late May of 1999, a new report was issued. It came up with a figure of 70 kilometers (43 miles) per second per megaparsec, plus or minus 7. At its lowest level, then, of 63 kilometers (39 miles), the age of the oldest stars in the Milky Way might just squeeze into the larger picture, especially because some other studies had lowered the ages of those stars. The leader of the team, Wendy Freedman of the Carnegie Observatories in

Pasadena, California (we've come a long way since the days of Henrietta Leavitt in terms of women's place in the field of astronomy), stated, "After all these years, we are finally entering the era of precision cosmology."

Those words were spoken on May 25, 1999. On June 1 came a different study, using different methods, announced at a meeting of the American Astronomical Society in Chicago. Its conclusions suggested problems with all previous galactic measurements. This study, based on radio astronomy and carried out by the Very Long Baseline Array of radio telescopes, measured the distance from the Earth to a galaxy 23.5 million light-years away in the constellation Ursa Major (the "Great Bear"). The Very Long Baseline Array consists of 10 identical dish antennae, each 82 feet (25 meters) in diameter. Used in concert, as they were in this study, they give a view equivalent to that of a telescope with a 5,000-mile (8,045-kilometer) diameter.

The measurements to Ursa Major suggested that the universe is 15 percent smaller than previously thought, and thus 15 percent younger as well. The galaxy studied, NGC 4258, was described by team member James Moran of Harvard as "nature's gift to radio astronomy," because of the presence of *masers,* a form of very strong radio emissions. While claimed to be the most accurate measurements of their kind ever made, the results again suggest a universe younger than the oldest stars in the Milky Way.

Something is wrong. Perhaps the redshift measurements of Cepheid stars, even though they go back to the 1920s, are (and have always been) off-base. It may be that the radio telescope measurements are based on wayward assumptions. And, although almost no one in the field of astronomy wants to say it openly, it could be that neither method of measuring distances is correct. The Big Bang theory itself could be the problem. There might indeed be an antigravity force, or some other undetected cosmological principle at work, or an unknown and as yet unimagined key to the problem. In any case, the most sophisticated efforts of the world's most eminent cosmologists are not producing answers that are in agreement, and until the measurements coalesce, the age of the universe will remain unknown.

But then, it should be remembered that as recently as the early 1920s, everyone knew that the entire universe consisted of the Milky Way alone.

⚛ To investigate further

Ferris, Timothy. *The Whole Shebang*. New York: Simon & Schuster, 1997. Ferris offers solid material on the age of the universe, although the problems currently afflicting the field developed after this book was published.

Thuan, Trinh Xuan. *The Secret Melody*. New York: Oxford University Press, 1995. Thuan's clarity and poetic gift enhance his treatment of this subject in particular.

Boslough, John. *Masters of Time*. Reading, MA: Addison-Wesley, 1992. This critical look at the problems and conflicts that have arisen in the field of cosmology draws engagingly on many interviews with top scientists.

Christianson, Gale E. *Edwin Hubble*. New York: Farrar, Straus and Giroux, 1995. This fine biography of Hubble clearly explains the development of his revolutionary concepts while also painting a vivid portrait of this colorful man and his rivalry with Harlow Shapley and others.

Hawking, Stephen. *A Brief History of Time* (10th anniversary edition). New York: Bantam, Doubleday, Dell, 1998. The original 1988 edition of this book by the legendary physicist was a huge international best-seller (perhaps more displayed than read). This updated and expanded edition adds new information and clarifies some points.

Note: For those who wish to keep up with the ongoing debates concerning the age of the universe, the *New York Times* provides excellent coverage, particularly in the articles on cosmology by John Noble Wilford and Malcolm Browne.

Chapter 19

Are There Multiple Universes?

Sometimes science fiction gets there first.

Space travel, courtesy of Jules Verne in his first novel *From the Earth to the Moon,* published in 1865, and its 1869 sequel, *Round the Moon,* became a popular concept a full century before *Apollo 11* landed on the moon, and 40 years before the Wright brothers managed to stay aloft in Earth's air at Kitty Hawk. Verne even had his spaceship taking off from Florida and splashing down in the Pacific a mere 2½ miles (4 kilometers) from where *Apollo 9* landed, as astronaut Frank Borman took the trouble to note in a letter to Verne's grandson following that flight. Among famous later examples of this kind of prescience, Cleve Cartmill perhaps takes first prize for his story "Deadline," which appeared in *Astounding Science Fiction* in the summer of 1944. The scientists in the story were involved in research very similar to what was actually being undertaken at the time by the physicists developing the atomic bomb. What really rang alarm bells in Washington was the fact that Cartmill had given the secret project in his story the code name "Hudson River Project." This was all too close to the real top-secret Manhattan Project, and both Cartmill and the magazine's editor, John W. Campbell, were grilled at length by the FBI. The pair finally convinced their questioners that it was "just a story," based on publicly available concepts that had been around for more than a decade.

A 1952 novel by Jack Williamson, however, is in its way an even more extraordinary example. Verne was extremely good at fleshing out ideas scientifically—he actually got the escape velocity right—but even the ancient Greeks had fantasized about flights to the moon, and the possibility of harnessing the energy of the atom had been a subject of debate for several decades when Cleve Cartmill took up the subject—H. G. Wells had coined the term "atomic bombs" in his 1913 novel *The World Set Free.* Williamson did something altogether different, however, in his 1952 book, *The Legion of Time:* He presciently foretold an event in *theoretical* science.

Williamson was one of the most imaginative of post–World War II science-fiction writers—far too much so for those who preferred their science fiction to be based on "hard science." In fact, he had a very good scientific background, but he was likely to take the slightest hint of a possibility and stretch it as far as he could. That's what he did in *The Legion of Time,* a story about traveling back and forth between parallel words, or universes. Science writer John Gribbin, in his 1984 book on quantum physics, *In Search of Schrödinger's Cat,* commented, "As far as I have been able to trace, this was the first time, in fact or fiction, that the concept of parallel worlds, later to become the many-worlds interpretation of quantum mechanics, appeared in print."

"Geodesics have an infinite proliferation of possible branches, at the whim of subatomic indeterminism," Williamson wrote in partial explanation of what was going on. Gibberish? Not at all. Gribbin pointed out that the physicist Hugh Everett, in his famous doctoral thesis on the subject written in 1957, "couldn't put it any more successfully, though he did put it on a secure mathematical footing." Everett caused a considerable ruckus with this thesis. In it he proposed the possibility that the universe continually "split" as it evolved, creating an infinite number of universes. It is not correct, however, to think of these universes as being parallel to one another. Instead each would fork off from the previous one, and another off of the new one. This carries the idea of "the road not taken" to its ultimate conclusion. In one universe (the one we know) Lincoln would be shot by John Wilkes Booth; in another, the wound would not be fatal; in still another, it would not have been fired at all; in still another, many forks further down the "road," neither Lincoln nor Booth would exist. There would be universes in which subatomic events had precluded the very existence of the United States, and others in which they precluded the evolution of the human race.

Even those who accept this possibility are willing to grant its power to boggle the mind. A much-quoted statement from physicist Bryce DeWitt, a proponent of the theory, puts it bluntly: "Every quantum transition taking place on every star, in every galaxy, in every remote corner of the universe is splitting our local world on Earth into myriads of copies of itself. I still recall vividly the shock I experienced on first encountering this multi-world concept." Common sense, indeed our very sense of the "real," can rebel at this idea, and for that reason it is disliked by many physicists. Even those who do not accept it, however, admit that there is nothing wrong with the mathematics that support it—they are right in line with other less dismaying interpretations of quantum theory. As we saw in chapter 15, new experimental studies are suggesting that quantum physics can work in the concrete world we know, as well as at the subatomic level—maybe.

Part of the resistance to the many-worlds theory derives from the fact that it is complicated and seemingly untestable—by definition, it would seem, there can be no communication between multiple universes, making tests of their existence impossible. It is worth recalling that the inflation theory about what happened just after the Big Bang is complicated and untestable also, yet physicists rushed to embrace that idea. Why not multiple universes? The reason for the difference in popularity between these two "far-out" concepts is that inflation solved a problem that was giving cosmologists fits. Thus they were willing to accept the untestable nature of that theory. However valid the multiple-universe theory may be as a mathematical implication of quantum mechanics, it solves no problem—it just creates new ones and is thus better ignored. The reader who feels that such a stance is all too convenient has company—proponents of the multiple-universe theory keep saying the same thing. Convenience can cut both ways, of course. It is convenient to cosmologists to accept inflation theory, but equally convenient to reject the many-worlds theory. John Wheeler, who oversaw the construction of the hydrogen bomb and gave black holes their name, was the teacher and mentor of Hugh Everett, and Wheeler contributed ideas to Everett's development of the many-worlds theory, but in the end he turned against it. The reason Wheeler gave was that it "required too much metaphysical baggage to carry around."

Wheeler himself worked toward conclusions that also resulted in multiple universes, although of a different kind. In his view, the (or rather a) universe expands to a certain point and then starts contracting. The contraction ultimately reaches a point where both the density and the temperature become infinite—and a new Big Bang inevitably occurs. However, each succeeding universe in this endless cycle will be different from the one before it—if the paths taken by even a few subatomic particles in the new universe are different from those that were followed in the previous universe, everything will be at least slightly different this time around, and everything could be almost unimaginably different, producing a new universe that has entirely different physical laws than ours does. Newton's gravity and Einstein's relativity would not apply. Indeed, as Trinh Xuan Thuan wrote in *The Secret Melody*, "Most of these cycles will not have the conditions necessary for the emergence of intelligence. By chance, our cycle happens to have the required conditions. . . . Wheeler has substituted an infinite succession of universes for Everett's frenzied duplication of universes, but the idea remains the same: an infinite number of universes, where the physical constants, the initial conditions, and even the physical laws may vary at random. Once again, these universes are completely unconnected with one another." Thuan, writing in the mid-1990s, also noted that Wheeler's cyclical theory had an "even weaker" scientific foundation than Everett's forking universes for several reasons, one of them being that there was no proof that the universe contained enough matter to

John Archibald Wheeler, who led the development of the hydrogen bomb and gave black holes their name, has proposed that the universe will eventually collapse back on itself and then explode in a new Big Bang, which will result in a universe with entirely different physical laws—an ultimate extension of quantum probability. Courtesy the American Institute of Physics, Emilio Segré Visual Archives.

cause it to collapse back on itself. That's important because the most recent astronomical evidence has supported the opposing view that the universe will go on expanding forever.

Stephen Hawking has taken ideas similar to Everett's in a third direction, which still implies multiple universes but in a way that is more acceptable to some cosmologists. As Michio Kaku pointed out in his 1994 book *Hyperspace,* Hawking started out "as a pure classical relativist rather than a quantum theorist." In other words, Einstein's theory of relativity and not Heisenberg's uncertainty principle informed his earlier work. Over the years, however, Hawking became convinced that only quantum theory could provide the "grand unified theory" he and other physicists have sought in recent years—a theory that would reconcile the quantum world with Newton and Einstein.

Quantum theory assumes a wave function that contains every possible future state of a given particle. Hawking decided to treat the entire universe as though it were a quantum particle; because such a particle has an infinite set of possible states, the concept of the universe having a wave function implies an infinite set of possible universes. The wave function seems to be partial to our own universe (or we would not be here to think about such things, to act as observers of the wave function), while most other universes are dead ones. It remains possible, however, that somewhere else the infinite possibilities inherent in the wave function have produced another universe even more "favored" than our own. In that putative universe the questions we are still struggling with might already have been solved by beings far more intelligent than we are.

Like Everett's forking universes, Hawking's multiple universes are beyond counting, but many physicists are happier with the Hawking version because of one thing: They aren't part of other universes, but separate from them, each a discrete bubble. In Everett's equations, our own actions create new universes, in which we get split off into alternative realities, although at some point on the way to infinity, the new fork would not contain you or me or the physicist. The quantum particles in that universe would have diverged just enough to leave us out of it altogether. To consider the matter in a more mundane if nevertheless awful way, we could be snuffed out much sooner because in the forking universe the oncoming automobile with the drunk at the wheel would hit us instead of just missing us.

Another aspect to the concept of Everett's forking universes disturbs many people—physicists, plumbers, and bank tellers alike: It seems to obliterate free will. No matter what we do, a new universe will be created in which we do not do it. In one forking universe or another, all possible outcomes to our every action will exist. Does it really matter, then, what we do? Our choices, for good or ill, no longer count. Physicists can find this idea particularly annoying—after all, their lives are spent trying to figure out exactly how things work. How deflating it would be if the answer to that quest was a universe in which all answers had equal status in their separate realities. Why bother?

On the other hand, the forking-universe theory can look more enticing to those who aren't happy with the way their lives have gone. How nice to know that in another universe, you did become a doctor instead of flunking out of medical school, did persuade your first true love to marry you instead of that phony other suitor, and did become a best-selling author instead of having an attic full of unsold manuscripts. Somewhere the football was caught, the soufflé did rise, the guy smiled back, the raise was granted. It's not a good idea to get carried away with this, though, because the next day could prove to be a nightmare, even in that alternative reality.

So far, we've looked at multiple-universe theories that have been proposed by eminent scientists, backed up by mathematical equations that physicists take seriously even when they don't like the implications of them: Everett's forking universes, Hawking's bubble universes, and Wheeler's endlessly reformulated universes that expand, contract, and are reborn in a Big Bang that makes everything different. But there is still another kind of multiple universe, which no one has backed up with mathematics, but which is not necessarily ruled out by quantum theory.

In the 1930s, Henry Hasse wrote a science-fiction story called "He Who Shrank." It had a profound effect on the young Isaac Asimov, who later included it in the anthology *Before the Golden Age*. In it, a scientist wishing to explore molecular structure concocts a mixture that will shrink a person down to molecular size—and then persuades his assistant to try it first. This part of the story is difficult to take seriously, pushing the elixir of Dr. Jekyll to its absolute limits. Hasse has a splendid payoff in store, however. The story is told in the first person by the assistant, who not only shrinks but keeps shrinking, down through one universe after another, over eons of time. At last, after several adventures among the bizarre inhabitants of other universes, the immortal traveler shrinks sufficiently so that he finds himself in our solar system, and eventually descends, a towering giant, into Lake Erie, thoroughly alarming the citizens of Cleveland. When he has shrunk to nearly human size, he seeks out a scientist and writer, hypnotizes him and tells his incredible story, which the scientist records in longhand while in this trance. Becoming conscious again, the scientist sees his visitor disappear into the very paper he has been writing on. He subsequently makes the visitor's story known to the world.

Without ever saying anything about the subject, Hasse manages to get across the idea that our world, our galaxy, our universe is merely one of the molecules in a tabletop in another, far larger universe, and that the very grains of sand on the shores of Lake Erie must also contain their own entire universes. Meanwhile, the poor hubristic scientist is shrinking down through layers and layers of universes, each tinier than the one before but each with its own complete cosmos.

This is just a story, a very clever one that gets across a mind-boggling idea through simple means—just a science-fiction story. Recall, however, that Jack Williamson's *The Legion of Time* was just a story, too, and that novel's multiple worlds were soon to be backed up by a brilliant mathematical proof that forced the world's greatest physicists to deal with its implications.

Are there multiple universes? There are great scientists who think there must be, but because we are shut off from them and cannot communicate with them, different physicists are free to construct numerous mathematically plausible scenarios of what they might be like. Because the implications of such universes are so unsettling, and they involve so much "metaphysical baggage," as John Wheeler said, quite a lot of physicists think the exploration of such ideas is essentially a waste of time. Leave that stuff to the philosophers and science-fiction writers, they say. But Hawking, Everett, and others have felt that unless such possibilities are taken into account and dealt with, the answers to questions that have far more pertinence to our own circumscribed existence will never be found. Even beyond that, there are the words of the philosopher St. Albertus Magnus: "Do there exist many worlds, or is there but a single world? This is one of the most exalted questions in the study of Nature." There will undoubtedly be those who will keep right on asking it, no matter how strange the answers may appear to be.

✳ To investigate further

Kaku, Michio. *Hyperspace.* New York: Oxford University Press, 1994. A theoretical physicist himself, Kaku is completely at home with such subjects as parallel universes and time warps. He also writes with admirable clarity and has a gift for using biographical details and references from other fields to engage the reader. He also has a sense of humor, a valuable asset in making a novice time traveler feel at ease.

Thorne, Kip S. *Black Holes and Time Warps.* New York: Norton, 1994. Thorne is a bit wary when it comes to the subject of parallel universes, but his long relationship with John Wheeler, as both student and colleague, illuminates the way physicists deal with one another when new theories create problems.

Berman, Bob. *Secrets of the Night Sky.* New York: Morrow, 1995. An astronomer and columnist for *Discover* magazine, Berman covers a host of topics in this book, which includes a chapter on other universes. His rather flip style will amuse some readers and irritate others.

Williamson, Jack. *The Legion of Time.* New York: Pyramid Books, 1952. This legendary science-fiction novel, first published in 1952 and reissued in 1967, is available on the World Wide Web from such sources as alibris.com and BookFinder.com. First editions of the 1952 original can run to over $100, but the 1967 reprint can be found at more reasonable prices.

Chapter 20

How Many Dimensions Are There?

Back in the 1950s, Hollywood, terrified that it would lose the attention of the nation's moviegoers to their television sets, came up with the three-dimensional movie. There we sat with cardboard glasses on our faces watching dreadful movies such as *Bwana Devil*. Boy, what a thrill. Not to be outdone, Howard Johnson's, then the country's premier fast-food chain, came up with a hamburger it called the "3-D," with two patties between three layers of bun—a concept that is still very much with us. In the twenty-first century, however, both Hollywood and the fast-food chains face a tall order. According to superstring theory, there are 10 dimensions, possibly 26, and how you turn that into a hamburger is anyone's guess.

The human race had managed quite nicely with the three spatial dimensions in which we live our lives until Einstein came along and gave us a fourth: time. Actually, this was not terribly difficult for the ordinary person to understand. If you agreed to meet a new friend at her office before taking in a showing of *Bwana Devil*, she would inform you that the building was located at the corner of Chestnut and King Street, say, and that you should come to the third floor. That takes in the right-left and back-forward dimensions of space in terms of the street corner, and the up-down aspect in terms of the floor designation. In addition, your new friend set a time, maybe 5:15, and that is another element of location. In terms of relativity, all actions take place not only in the three spatial dimensions but also in the fourth dimension of time. When you put all four together you get Einstein's spacetime.

In 1919, shortly after the general theory of relativity was confirmed by Arthur Eddington's observations of Mercury during an eclipse of the Sun, Einstein received a letter from a Polish mathematician who was as obscure as Einstein himself had been before 1905. The mathematician, Theodor Kaluza, put forward the idea that the universe might have more than three spatial dimensions. Kaluza's reasoning involved the possibility that there might be a curled-up dimension too small to be seen. Attempts to explain this curled-up dimension tend to be tortuous, even when accompanied by illustrations, precisely because it is impossible in our large, macro three-dimensional world to represent anything in more than three dimensions, even in a sculptured object and, on the page of a book, in more than two dimensions. Brian Greene, a physicist who not only understands this field of physics but also has contributed importantly to it, spends several pages in his 1999 book *The Elegant Universe* using an analogy based on a garden hose stretched across a canyon, with an ant crawling on it. This actually gets the idea across in the end, but suffice it to say here that the hose takes on quite different appearances to a person viewing it with or without binoculars, is different still to the ant, and contains within it a curled-up space that is unseen to all.

"Unseen to all" is the operative phrase here. The extra dimension that Kaluza suggested to Einstein—and the increasing numbers of extra dimensions that have been added since the early 1980s—cannot be observed with any instrument we have. Mathematically, however, the assumption of its existence produced astonishing results. What initially caught Einstein's attention was that the relativity formulas that Kaluza worked out using an extra dimension led inexorably to equations that James Clerk Maxwell had used to describe the electromagnetic force in the 1880s. Einstein's own work had developed out of Maxwell's, but it was only with the addition of an additional dimension that electromagnetism and relativity were fully united. Einstein waxed hot and lukewarm about Kaluza's ideas for two years and then agreed to see that Kaluza's paper was published. Kaluza's ideas were then augmented by Swedish mathematician Oskar Klein. Experiments intended to prove the theory ran into serious problems, however, and the idea was shunted aside.

It was not until the 1970s that Kaluza's ideas resurfaced in connection with string theory. The first glimmerings of this new theory were unveiled by accident by Gabriel Veneziano, a young research intern at CERN, the accelerator laboratory in Geneva, Switzerland. Veneziano was working on problems connected with the strong nuclear force. Leafing through a math book, his eye was caught by an esoteric nineteenth-century mathematical function devised by mathematician Leonhard Euler. Veneziano saw that the Euler beta function, as it is called, seemed to describe many of the strong reactions among elementary particles. This was the starting point for an entirely new way of looking at the universe. Quantum physics was running into all kind of problems at the time, and younger physicists quickly became interested in this fresh theoretical direction. Bit by bit, additional aspects of what would become string theory began to emerge throughout the 1970s. There appeared to be a lack of internal consistency in the material, however, and it was not until John Schwarz of the California Institute of Technology and Michael Green of Queen Mary's College in London were able to show in 1984 that self-consistency was attainable that string theory truly took off.

What are *strings*? They are entities that vibrate throughout the universe, everywhere, so infinitesimal that it takes 10 million billion of them to make up a quark—which is itself so small we can only infer its existence from experiments. We are going down to a level that underlies the subatomic world of quantum physics, a realm of activity so infinitesimal that the word "micro" seems utterly inadequate. Some readers may be reminded here of the medieval debates about how many angels can dance on the head of a pin, or of the Henry Hasse short story recounted in chapter 19, in which a scientist disappears into a tabletop and reappears as a giant in Lake Erie. Many eminent physicists had the same initial reaction—and some are still very dubious.

String theory does have something important going for it, however. The intractable problem of how to work the gravitational force into quantum physics disappears. It doesn't simply provide a formula that unites the two, either. String theory insists that gravity must exist. Indeed, Edward Witten, the acknowledged leader of the string-theory contingent, goes further: "String theory has the remarkable property of *predicting gravity*." Brian Greene explains what is meant by this: "Both Newton and Einstein developed theories of gravity because their observations of the world clearly showed them that gravity exists, and that, therefore, it required an accurate and consistent explanation. On the contrary, a scientist studying string theory—even if he or she was completely unaware of general relativity—would be inexorably led to it by the string framework."

Even Greene, a major backer of string theory, sees a problem here, noting that because we already know all about gravity, string theory's prediction of it is more of a "postdiction." Given that the mathematics used in elucidating string theory are new in themselves, and that mathematics in general can be forced to desired conclusions (corporations and governments do it all the time), there was considerable resistance to Witten's sense of triumph in this regard. Even so, the fact that string theory united gravity with the other three fundamental forces (the electromagnetic and the strong and weak nuclear forces) with relative ease certainly put it one up on quantum theory.

Still, there is the matter of those extra dimensions. String theory, it was quickly apparent, demanded the existence of an additional 6 spatial dimensions beyond the 3 we are conversant with in our daily lives. When Einstein's dimension of time is added, that produced a total of 10 dimensions—a nice round number. These additional dimensions, like the vibrating subsubatomic strings, were of course invisible to us—and destined to stay that way until our technology catches up. Edward Witten has also said that string theory is twenty-first-century science that was discovered too soon for it to be proved with our existing means of investigation. This, too, can sound convenient, but it should be remembered that Charles Babbage had laid down all the fundamental laws of computing by 1830, but because he was stuck with the entirely inadequate technology of punch cards, his work was forgotten for more than a hundred years. Scientific theory often does get ahead of the level of technology available to either implement or prove it.

Nevertheless, it must be asked what things look like in this infinitesimal world with 10 dimensions. The string theorists have answers to that, up to a point. Throughout *The Elegant Universe,* there are illustrations that attempt to depict what are called Calabi-Yau spaces. These are named in honor of two mathematicians, Eugenio Calabai and Shing-Tung Yau, whose research, even though not related to string theory, helped define such spaces. The pictures, as Greene keeps pointing out, are only approximations because they are representing a 6-dimensional shape on a 2-dimensional page. Basically, they look like someone took one of M. C. Escher's well-known drawings of stairways going nowhere and rolled it up into a kind of ball of yarn. The ball shape is no accident. The extra dimensions posited by string theory are curled up and thus as difficult to see as the inside of Greene's hose stretched across the canyon with the ant crawling on it. These 6-dimensional spaces exist within the 3 dimensions we know and can see. Although we would be totally disoriented inside such a space, the infinitesimal vibrating strings that theoretically underlie everything in the universe are completely at home in them.

In fact, according to string theory, it is the ways in which the infinitesimal vibrating strings move through these extra 6 dimensions that determines the masses of particles and the charges of forces on the subatomic level, which in turn affect what happens in the real world we inhabit. The extra dimensions are not arbitrary, in other words. They are necessary to the particular "resonances" that the strings produce, just as in our macroworld the shape of a violin and the wood it is made from create a slightly different resonance when the strings are plucked. Of course, in the 10-dimensional world of string theory, the variety of resonances that would be produced are vastly increased, to the point that the resonances are capable of running an orderly universe.

Another version of string theory involves not just 10 dimensions, but also an additional 26. According to this approach, there are two types of vibrations, one type moving clockwise through

10 dimensions, and another that moves counterclockwise through 26. (The very use of the words *clockwise* and *counterclockwise* in connection with 10 dimensions, let alone 26, can be seen as a limitation of language, or as a sign that we have gone down Lewis Carroll's rabbit hole into Wonderland here.) Michio Kaku, another string theorist, explains in his 1994 book *Hyperspace* that this "heterotic string owes its name to the fact that the clockwise and counterclockwise vibrations live in two different dimensions but are combined to produce a single superstring theory. That is why it is named after the Greek word *heterosis,* which means 'hybrid vigor.'" To superstring theorists, the beauty of all these dimensions is that they create "room enough to explain all the symmetries found in both Einstein's theory and quantum theory." The phrase "room enough" is extremely important. What excites many physicists about string theory is that, mathematically speaking, "the laws of physics simplify in higher dimensions," as Kaku puts it. Crudely speaking, it is like adding several new file cabinets to the office—suddenly there is space enough to contain much more data in a more organized way.

The fact that none of these concepts are testable with current technology, however, bothered a great many physicists, as string theory was being developed in the 1980s—although others found it tremendously exciting. Nobel Prize winners were to be found on both sides of the debate. Murray Gell-Mann, who predicted the existence of the quark and named it, went on record as saying that he thought some version of string theory would ultimately trump all other theories. Taking the opposite view, Sheldon Glashow joined forces with a Harvard colleague to decry "magical coincidences, miraculous cancellations and relations among seemingly unrelated (and possibly undiscovered) fields of mathematics."

As quantum physicists continued to bang their heads against the gravity conundrum, and the string theorists calmed down a little and admitted some problems, an uneasy peace descended on the world of physics. At least string theory seemed to be going somewhere, while quantum theory kept encountering the same old problems. Nonetheless, although string theory dealt better with the gravity problem and a number of others, it had its own drawbacks. As Timothy Ferris points out in *The Whole Shebang,* the proliferation of supposed subatomic particles, now a zoo of more than 300, which had bedeviled quantum theory for some time was now affecting superstring theory. (The names were equally cute, as well, including the squark and the sneutrinio). In addition, it has not answered the question of how the other six dimensions "curled up" in the first place. As Ferris wrote, "A field theory of strings should derive the masses of the proton and other particles, but no such theory has yet been devised." Michio Kaku says, very directly, that no one is smart enough to solve the field-theory problem. This is another variation on Witten's theme that string theory is twenty-first-century physics accidentally plopped down in the twentieth century. (It is interesting to note that Curt Suplee's *Physics in the 20th Century,* published under the auspices of the American Physical Society and the American Institute of Physics, resolutely ignores string theory.)

Kaku admits to another difficulty. No one has any idea why superstring mathematics work only in either 10 dimensions or 26 dimensions. The equations fall apart otherwise—which is why Sheldon Glashow refers to magic numbers. To make matters worse, the idea has been put forward that it may be necessary to deal with 11 dimensions instead of 10, adding not another spatial dimension but a second time dimension to the one proposed by Einstein.

A number of top physicists have come to the conclusion that string theory is either the complete answer to all the mysteries of physics, one that will ultimately unite Newton's and Einstein's universes with quantum theory—or it will turn out to be utterly false, a wild-goose chase of cosmic proportions. There have always been dead-end theories in physics. We hear about Aristotle's Earth-centered universe because that totally wrong idea dominated human thought for so long, but most incorrect theories disappear into the wastebasket before the general public hears anything about them. If string theory does turn out to be wrong, many very important physicists are going to wish they could go hide someplace in an invisible dimension. Of course, it could be that none of them will still be around to be embarrassed. Both the technology and the kind of new mathematics necessary to solving the mysteries of string theory may well lie decades in the future.

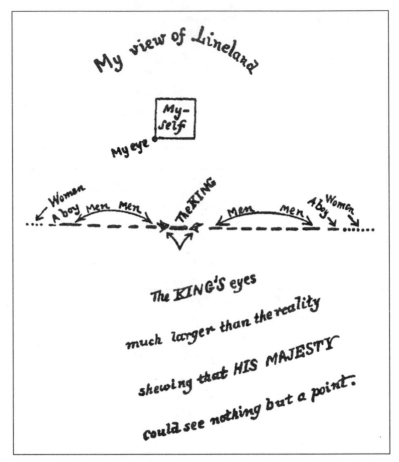

The KING'S eyes

much larger than the reality

shewing that HIS MAJESTY

could see nothing but a point.

This is one of the drawings made by Edwin A. Abbott for his 1884 satirical fantasy
Flatland, narrated by one A. Square, who inhabits a two-dimensional world, but
explores other dimensions, in this case the one-dimensional Lineland, and meets a
tragic fate when he suggests that a three-dimensional world might exist. The dedication
to Abbott's book reads, "To the Inhabitants of SPACE IN GENERAL And H.C. IN
PARTICULAR This Work is Dedicated By a Humble Native of Flatland, In the Hope
that Even as he was Initiated into the Mysteries of THREE Dimensions Having been
previously conversant With ONLY TWO So the Citizens of that Celestial Region May
aspire yet higher and higher to the Secrets of FOUR FIVE OR EVEN SIX Dimensions
Thereby contributing To the Enlargement of THE IMAGINATION And the possible
Development Of that most rare and excellent Gift of MODESTY Among the Superior
Races Of SOLID HUMANITY."

⚛ To investigate further

Greene, Brian. *The Elegant Universe.* New York: Norton, 1999. This is the most complete presentation of string theory yet published for the general public. Greene has been on the front lines in this field, and he certainly knows the territory inside out. He writes well and is honest about the problems involved almost to a fault: In anticipating the reader's skepticism, he sometimes undercuts his own arguments. Although lucid and fluent, this is not a book for the casual reader.

Kaku, Michio. *Hyperspace.* New York: Oxford University Press, 1994. Subtitled "A Scientific Odyssey through Parallel Universes, Time Warps, and the Tenth Dimension," Kaku's book is less dense than Greene's and is more relaxed. The author has a gift for bringing in ideas from other fields (including literature), which illuminate scientific ideas and add some fun to the mix.

Ferris, Timothy. *The Whole Shebang.* New York: Simon & Schuster, 1997. Ferris, as noted before, is one of our very best science popularizers, probably the best with respect to physics. He handles string theory a bit gingerly (I sympathize entirely), but the basic facts are here.

Abbott, Edwin A. *Flatland: A Romance of Many Dimensions.* Mineola, NY: Dover, 1992. Originally published in 1884, this classic work of imaginative fiction concerns life in a two-dimensional universe. Its delights have only been enhanced by the discoveries of twentieth-century physics. This classy little unabridged Dover edition only costs a buck.

Chapter 21

How Will the Universe End?

Our sun, roughly 4.6 billion years old, is about halfway through its life as a star. It is a commonplace star—there are billions like it in the universe. Such stars are constantly dying, and others like it are constantly being born. Having observed the corpses of similar stars born earlier in the history of the universe, we know in considerable detail what the death throes of our sun will be like. In about 4 billion years, our sun will exhaust the hydrogen that fuels its nuclear furnace. It will begin to contract, but then it will find a new lease on life: The helium nuclei within it will fuse in threes to form carbon-12, and this new source of fuel will last for another 2 billion years. While our sun will continue to live, the planet Earth will not. Burning this new kind of fuel will cause the size of our sun to increase 100 times over, turning Earth to a cinder that will be absorbed by the new red giant at the center of the solar system. Eventually, when the conversion of helium to carbon-12 has run its course, our sun will contract again, this time collapsing into a faint white dwarf. For several billion more years, the white dwarf will gradually cool down and eventually become a dead star known as a black dwarf.

There is a problem with this scenario, however, a mystery that might be called "The Case of the Missing Solar Neutrinos."

When Wolfgang Pauli first postulated the existence of the neutrino in 1931, it was because he needed a way to explain the fact that a small quantity of energy was absent from the electron produced by the radioactive decay of an atom. Because the laws of the conservation of energy demanded that the energy emitted by the atom and the energy carried away by the electron be equal, he decided there must be a "ghost particle" that stole away invisibly with the missing energy. Neutrinos were thieves of energy. It took more than two decades for the existence of the neutrino, which carries no charge, to be confirmed. There turned out to be three different kinds, or "flavors"—to use another of those cute quantum terms—of neutrino, which are identical aside from having different masses.

Neutrinos are emitted from the Sun in great quantities, a product of the nuclear fusion taking place in its core. They are extremely difficult to detect (as ghosts should be), but experiments of several different kinds have established without question that they are indeed streaming out of the Sun and passing through the Earth itself, as well as us, on their invisible journey outward into space. There are not enough of them, however. Depending on the techniques used to detect the neutrinos, a third to a half of the number that ought to be produced by the Sun are missing. Somehow, these thieves of energy are themselves being waylaid between the Sun and the Earth.

This problem has existed for 30 years. Because all the other evidence supporting the concept that the Sun is powered by nuclear fusion is so strong, the missing neutrinos are regarded as a puzzle that will eventually be solved through improved experiments, rather than as a real challenge to the prevailing model of solar activity. However, a few scientists with a creationist view of the universe, firmly opposed to the theory of evolution, have seized on the missing neutrinos to argue that the Sun is not powered by nuclear fusion at all and is thus far younger than the standard model holds. A much younger Sun would mean a much younger Earth, young enough to dispense with the whole notion of evolution. Their arguments have been thoroughly demolished by numerous mainstream scientists; the overwhelming evidence points to the Sun being indeed 4.6 million years old and about halfway through its life span, missing neutrinos or no.

Even before the death of our sun, our Milky Way galaxy will eat the dwarf galaxy known as the large Magellanic Cloud, and it will endure a major collision with the Andromeda galaxy. The large Magellanic Cloud, only 150,000 light-years away now, is slowing down and being drawn by gravitational force toward the Milky Way, which will devour it in about 3 billion years, adding a million stars to the Milky Way's bulk, which may prove useful when the Milky Way and Andromeda collide about 700 million years later. Space is vast, galaxies collide all the time, and surprisingly little damage is done. Sure, a few stars will bump into one another, with dire consequences for any planets they may harbor within their vicinity, but that happens only occasionally, and the collisions of galaxies are mere fender benders on the cosmic scale.

The larger question revolves around whether the universe is expanding or contracting. This is a fairly recent subject of contention. After all, it wasn't until Edwin Hubble's 1925 paper about "island universes" that we even knew there was more than one galaxy, our own Milky Way. When Einstein was developing the general theory of relativity, even he assumed that there was just one galaxy in the universe and that it was static. Because his formulas suggested that the (one galaxy) universe ought to be expanding, he applied his cosmological constant to prevent it from doing that. Once Hubble had demonstrated that there were innumerable galaxies, and that they were moving away from one another, expanding the size of the universe, Einstein threw out the cosmological constant and lamented that he hadn't trusted his own figures in the first place.

Soon, however, new arguments arose about the expansion of the universe. It might be expanding now, some cosmologists argued, but eventually it would stop doing that and would collapse back in on itself. As the Big Bang was taken more seriously in the second half of the twentieth century and came to be generally accepted in the early 1980s, many scientists were convinced that the propulsive outward energy created by the Big Bang would eventually diminish, the expansion would slow down, stop, and then go into reverse, with all the stars and galaxies falling back

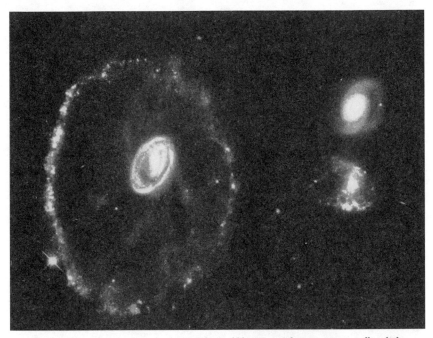

The Cartwheel Galaxy, as photographed by the Hubble Space Telescope, is 500 million light-years away in the constellation Sculptor. Its unusual configuration resulted from the collision of two galaxies, which created a new one larger than our own Milky Way. The bright splotches to the right of the picture are smaller galaxies, but it is not known which of the two was involved in the collision. Eventually the Cartwheel Galaxy will settle once again into a spiral, a process that can be seen taking place in the faint "spokes" starting to emanate from the central core. Courtesy NASA (Kirk Borne, Space Telescope Science Institute).

together again, eventually to crunch together in a cataclysmic pig-pile sometimes referred to as the Big Crunch. The residue of that cataclysm would be so dense and so hot that it would eventually become a pinpoint once more, containing all the matter and energy in the universe—ready to explode all over again in another Big Bang. The strongest proponent of this view was American physicist John Wheeler. According to his theory, this whole process goes on into infinity, with each new Big Bang creating a universe with entirely different laws because the slightest deviation of a single electron at quantum levels would be enough to change the nature of everything (see chapter 19).

This cyclical pattern has strong philosophical appeal to many cosmologists, and the mathematics are perfectly sound. The myth of the phoenix that rises from its own ashes is deeply embedded, one way or another, in most religions, which for a time gave Wheeler's views a psychological advantage in debates about the end of the universe. Rebirth is a seductive idea, even at the cosmic level.

Another school of thought takes the position that while this cycle is all very nice, it doesn't conform to what we are observing, and that the actual end of the universe is a somewhat bleaker proposition. In this view, expansion will continue forever. (It should be said here that the universe is not believed to be expanding into anything except utter emptiness, a conception that tends to bother the ordinary person but not cosmologists.) As the galaxies get farther and farther apart from one another, the collisions that foster the birth of new ones will diminish. The cold vacuum between galaxies will become ever greater, and the stars within those galaxies will gradually burn up all their fuel, just as our sun will. Stars larger than 1.4 times the size of our sun will have a far more violent and extended demise, but they, too, will ultimately use up all their energy.

After 1,000 billion years (we're now at 8 to 15 billion, depending on who's counting, as detailed in chapter 18), there will be almost nothing but dead stars and black holes in a darkened universe. Even those, however, because of the endless pull of gravity, will have another go at a cosmic fireworks display, some billion billion years after the Big Bang. There will be light again for about a billion years, less than a quarter of the time Earth has existed, and then, over a period of unimaginable time the universe will become completely dark and cold, as even the remaining black holes evaporate. How long will this process take? As Trinh Xuan Thuan puts it, "To write out such a number I would have to follow the figure 1 with as many zeroes as there are hydrogen atoms in all the hundreds of billions of galaxies in the observable universe." All that will ultimately be left is radiation and virtual quantum particles that will wink in and out of existence instantaneously forever and ever.

New evidence reported in the year 2000 suggested that the universe is expanding at a much faster rate than previously thought, perhaps reducing the time scale described here. In addition, all kinds of possibilities could change that view. As we have seen in the course of this book, even the present age of the universe is so open to question that the very techniques being used to measure the time frame are now suspect. Quantum physics is only beginning to reveal the bizarre mysteries of the subatomic world to us. An electron can be in two places at once, and seemingly electrons can communicate with one another at a distance, informing one another how to behave when a guest is present to observe them. As the new millennium began, there was a good deal of self-congratulation in the air about the vast strides forward that science made in the twentieth century, in field after field. In a hundred years, the human race gained more knowledge about the universe and its constituent parts, large and small, from galaxies to genes, than in all of previous history combined. Certainly that calls for a modest degree of celebration, but it is important to remember how much we don't know:

The Big Bang is a matter of theory, much of it untestable.

We have only the vaguest ideas about how life on Earth got started.

We believe we know what caused the extinction of the dinosaurs, at last, but what about the other major periods of extinction?

The inside of the Earth is much better understood, but we still can't predict earthquakes in any helpful way.

Some of the factors that contribute to ice ages are understood, but their relationship to one another remains extremely foggy.

The debate over whether dinosaurs were warm-blooded or cold-blooded has gotten hotter, not colder.

The record of the evolution of our own kind remains full of gaps.

Humankind's sudden leap to civilization remains a deep puzzle.

We haven't even figured out how we acquire language.

Some scientists suspect that dolphins have an intelligence almost equal to ours and could teach us a great deal—if only we could learn to communicate with them.

Bird migration is still a wonder to us—perhaps even satisfyingly so.

It is impossible to know whether the grass is really green to anyone but us.

The mysterious astronomical and calendrical achievements of the Maya suggest that degrees of knowledge depend on what is being looked for.

Scientists have failed to integrate the force of gravity with the other three fundamental forces.

Light seems to be a wave sometimes, a particle at others, and the dividing line is still theoretical.

Quantum physics is haunted by a cat that's both dead and alive.

It is now certain that black holes exist, but we don't really have any idea what goes on inside them.

The age of the universe is up for grabs.

The mathematical possibility that every move we make creates a new universe has turned Oz into a model of reality.

So many dimensions have now been postulated that the arrival of a fourth in the early twentieth century with Einstein's spacetime continuum seems almost quaint by comparison.

Still, we want to know how the universe might end. Kind of cheeky of us, considering the number of other unsolved mysteries of science, isn't it? Surely, that is a good part of the pleasure of being alive. We want to know everything, and we keep trying to find the answers.

✷ To investigate further

Livio, Mario, with foreword by Allan Sandage. *The Accelerating Universe: Infinite Expansion, the Cosmological Constant, and the Infinite Beauty of the Cosmos.* New York: John Wiley & Sons, 2000. Livio is one of the directors of research for the Hubble Space Telescope and is in the thick of the latest developments and conundrums. He also considers the human need to impose order on the universe and discusses how that need affects scientists. This is an eloquent and extremely well-received book.

Gribbin, John. *In Search of the Big Bang: The Life and Death of the Universe.* New York: Penguin, 1999. This much-revised new edition updates one of the first popular books on the subject.

Thuan, Trinh Xuan. *The Secret Melody.* New York: Oxford University Press, 1995. Thuan's discussion of the end of the universe is as clear and comprehensive a short treatment as you will find, and I drew on it extensively here.

Note: For a scenario that unites computers, religion, and the end of the universe in one neat package, readers can look to a classic of science fiction, Arthur C. Clarke's most famous short story, "The Nine Billion Names of God." It has been anthologized innumerable times, for good reason, and is as provocative a vision as anything cosmologists have ever managed to put forward.

Index